Aspects of Differential Geometry III

Synthesis Lectures on Mathematics and Statistics

Editor
Steven G. Krantz, *Washington University, St. Louis*

Aspects of Differential Geometry III
Esteban Calviño-Louzao, Eduardo García-Río, Peter Gilkey, JeongHyeong Park, and Ramón Vázquez-Lorenzo
2017

The Fundamentals of Analysis for Talented Freshmen
Peter M. Luthy, Guido L. Weiss, and Steven S. Xiao
2016

Aspects of Differential Geometry II
Peter Gilkey, JeongHyeong Park, Ramón Vázquez-Lorenzo
2015

Aspects of Differential Geometry I
Peter Gilkey, JeongHyeong Park, Ramón Vázquez-Lorenzo
2015

An Easy Path to Convex Analysis and Applications
Boris S. Mordukhovich and Nguyen Mau Nam
2013

Applications of Affine and Weyl Geometry
Eduardo García-Río, Peter Gilkey, Stana Nikčević, and Ramón Vázquez-Lorenzo
2013

Essentials of Applied Mathematics for Engineers and Scientists, Second Edition
Robert G. Watts
2012

Chaotic Maps: Dynamics, Fractals, and Rapid Fluctuations
Goong Chen and Yu Huang
2011

Matrices in Engineering Problems
Marvin J. Tobias
2011

The Integral: A Crux for Analysis
Steven G. Krantz
2011

Statistics is Easy! Second Edition
Dennis Shasha and Manda Wilson
2010

Lectures on Financial Mathematics: Discrete Asset Pricing
Greg Anderson and Alec N. Kercheval
2010

Jordan Canonical Form: Theory and Practice
Steven H. Weintraub
2009

The Geometry of Walker Manifolds
Miguel Brozos-Vázquez, Eduardo García-Río, Peter Gilkey, Stana Nikčević, and Ramón Vázquez-Lorenzo
2009

An Introduction to Multivariable Mathematics
Leon Simon
2008

Jordan Canonical Form: Application to Differential Equations
Steven H. Weintraub
2008

Statistics is Easy!
Dennis Shasha and Manda Wilson
2008

A Gyrovector Space Approach to Hyperbolic Geometry
Abraham Albert Ungar
2008

© Springer Nature Switzerland AG 2022
Reprint of original edition © Morgan & Claypool 2017

All rights reserved. No part of this publication may be reproduced, stored in a retrieval system, or transmitted in any form or by any means—electronic, mechanical, photocopy, recording, or any other except for brief quotations in printed reviews, without the prior permission of the publisher.

Aspects of Differential Geometry III

Esteban Calviño-Louzao, Eduardo García-Río, Peter Gilkey, JeongHyeong Park, and Ramón Vázquez-Lorenzo

ISBN: 978-3-031-01282-2 paperback
ISBN: 978-3-031-02410-8 ebook

DOI 10.1007/978-3-031-02410-8

A Publication in the Springer series
SYNTHESIS LECTURES ON MATHEMATICS AND STATISTICS

Lecture #18
Series Editor: Steven G. Krantz, *Washington University, St. Louis*
Series ISSN
Print 1938-1743 Electronic 1938-1751

Aspects of Differential Geometry III

Esteban Calviño-Louzao
University of Santiago de Compostela, Spain

Eduardo García-Río
University of Santiago de Compostela, Spain

Peter Gilkey
University of Oregon

JeongHyeong Park
SungKyunkan University, Korea

Ramón Vázquez-Lorenzo
University of Santiago de Compostela, Spain

SYNTHESIS LECTURES ON MATHEMATICS AND STATISTICS #18

ABSTRACT

Differential Geometry is a wide field. We have chosen to concentrate upon certain aspects that are appropriate for an introduction to the subject; we have not attempted an encyclopedic treatment. Book III is aimed at the first-year graduate level but is certainly accessible to advanced undergraduates. It deals with invariance theory and discusses invariants both of Weyl and not of Weyl type; the Chern–Gauss–Bonnet formula is treated from this point of view. Homothety homogeneity, local homogeneity, stability theorems, and Walker geometry are discussed. Ricci solitons are presented in the contexts of Riemannian, Lorentzian, and affine geometry.

KEYWORDS

Chern–Gauss–Bonnet formula, G-structures, gradient Ricci solitons, homothety homogeneity, invariance theory, local homogeneity, pseudo-Kähler geometry, Ricci almost solitons, Ricci solitons, VSI manifolds, Walker geometry, Weyl invariants.

This book is dedicated to
Alison, Carmen, Fernanda, Hugo, Junmin, Junpyo, Luis, Manuel,
Montse, Rosalía, Sara, and Susana.

Contents

Preface

This three-volume series arose out of work by the authors over a number of years both in teaching various courses and also in their research endeavors. For technical reasons, the material is divided into three books and each book is largely self-sufficient. To facilitate cross references between the books, we have numbered the chapters of Book I from 1–3, the chapters of Book II from 4–8, and the chapters of the current volume from 9–11. A fourth volume dealing with affine surfaces is in preparation.

Chapter 9 deals with invariance theory. We begin with a brief review of affine and pseudo-Riemannian geometry to keep the current volume as self-contained as possible, a more complete treatment being given in the previous volumes. We discuss Weyl's first and second theorem of invariants and introduce the theory of universal curvature invariants. We discuss the Euler–Lagrange equations for the Pfaffian and derive the Chern–Gauss–Bonnet formula for a closed manifold in the indefinite setting from the corresponding result in the Riemannian context. We also treat the corresponding formula for manifolds with boundary. The Pfaffian is an unstable characteristic class and this instability manifests itself in the invariance theory. A similar treatment is given in the pseudo-Kähler setting to discuss the Euler–Lagrange equations defined by the Chern forms. A pseudo-Riemannian manifold is said to be VSI (vanishing scalar invariants) if all the Weyl scalar invariants vanish. A brief introduction to VSI geometry is presented as this is a rich source of examples. Since any Riemannian VSI manifold is necessarily flat, this subject is only relevant to the higher signature context. We present some examples and develop the elementary properties of generalized plane wave manifolds. We conclude the chapter by discussing some invariants which are not of Weyl type and which arise in various settings in affine geometry.

Chapter 10 deals with local homogeneity. The notion of homothety homogeneity and the homothety character are introduced. It is shown that if the manifold in question is not VSI and if the homothety character is non-trivial, then in fact the manifold is not homogeneous. We present a classification result and show that in the Riemannian setting such a manifold is necessarily incomplete. The situation is not as rigid in the Lorentzian setting where pp-waves support non-Killing homothety vector fields. We establish a stability result that lets us pass from the algebraic to the geometric level. We treat Walker Lorentzian manifolds both in the context of curvature homogeneity and in the context of homothety curvature homogeneity. We also examine locally homogeneous metric G-structures.

Chapter 11 treats the geometry of Ricci solitons and focuses on their classification under various geometric conditions. The pseudo-Riemannian setting permits the existence of many Ricci solitons that do not have a Riemannian analogue. This is caused, in part, by the existence of non-flat pseudo-Riemannian manifolds which support homothetic vector fields as discussed in Chapter 10. We examine the existence of Ricci solitons and Ricci almost solitons in the homogeneous Riemannian setting. Locally homogeneous gradient Ricci solitons in the Lorentzian context are treated and the structure of the Ricci operator discussed; in the non-steady case, the soliton is rigid in dimensions 3 and 4; in the steady case, a complete classification is given in dimension 3. Locally conformally flat gradient Ricci solitons are discussed in both the Riemannian and Lorentzian situation. Finally, self-dual gradient Ricci solitons are considered with special attention to their link to affine geometry, thus leading to many examples of strictly self-dual gradient Ricci solitons in neutral signature.

Acknowledgments

We have provided many images of famous mathematicians in this book; mathematics is created by real people and we think having such images makes this point more explicit. The older pictures are in the public domain. We are grateful to the Archives of the Mathematisches Forschungsinstitut Oberwolfach for permitting us to use many images from their archives; the use of this images was granted to us for the publication of this book only and their further reproduction is prohibited without their express permission. Some of the images provided to us by the MFO are from the collection of the Mathematische Gesellschaft Hamburg; again, the use of any of these images was granted to us for the publication of this book only and their further reproduction is prohibited without their express permission.

The research of the authors was supported by the Basic Science Research Program through the National Research Foundation of Korea (NRF) funded by the Ministry of Education (NRF-2016R1D1A1B03930449) and by Projects EM2014/009, MTM2013-41335-P, and MTM2016-75897-P (AEI/FEDER, UE). The assistance of Ekaterina Puffini of the Krill Institute of Technology has been invaluable. Wikipedia has been a useful guide to tracking down the original references and was a source of many of the older images that we have used that are in the public domain.

Esteban Calviño-Louzao, Eduardo García-Río, Peter Gilkey, JeongHyeong Park, and Ramón Vázquez-Lorenzo
May 2017

CHAPTER 9

Invariance Theory

The material in this chapter arose out of work by Brozos-Vázquez, García-Río and Gilkey [19, 20], Díaz-Ramos, García-Río and Nicolodi [56], Gilkey [73], Gilkey and Park [76], Gilkey, Park and Sekigawa [77], and Park [112]. In Section 9.1, we review the basic material concerning pseudo-Riemannian and affine geometry that we will need in this and subsequent chapters. In Section 9.2, we discuss Weyl's first and second theorem of invariants and introduce the theory of universal curvature identities. In Section 9.3, we discuss the Euler–Lagrange equations related to the Pfaffian for Gauss–Bonnet gravity, derive the Chern–Gauss–Bonnet formula for a closed manifold in indefinite signature from the corresponding result in the Riemannian setting, and treat the Chern–Gauss–Bonnet formula for manifolds with boundary. In Section 9.4, we turn our attention to the pseudo-Kähler setting and examine Euler–Lagrange equations defined by Chern forms which generalize the corresponding Euler–Lagrange equations for the Pfaffian discussed previously. In Section 9.5, we give some examples of *vanishing scalar invariant* (VSI) manifolds; these are pseudo-Riemannian manifolds all of whose scalar Weyl invariants vanish. We also develop some of the elementary properties of generalized plane wave manifolds. In Section 9.6, we discuss invariants which are not of Weyl type which arise in various settings in affine geometry.

9.1 REVIEW OF AFFINE AND RIEMANNIAN GEOMETRY

In this section, we introduce various concepts we will need subsequently. Section 9.1.1 treats affine geometry, Section 9.1.2 deals with Riemannian geometry, Section 9.1.3 treats the relation between the Levi–Civita connection and the metric, and Section 9.1.4 introduces various notions of homogeneity. Section 9.1.5 presents some examples of curvature homogeneous spaces which are not locally homogeneous. Section 9.1.6 deals with manifolds which are homothety homogeneous.

9.1.1 AFFINE GEOMETRY. Let $\vec{x} = (x^1, \dots, x^m)$ be a system of local coordinates on a smooth manifold M of dimension m. Set $\partial_{x^i} = \frac{\partial}{\partial x^i}$. Let ∇ be a connection on the tangent bundle TM of M. We adopt the *Einstein convention* and sum over repeated indices to expand $\nabla_{\partial_{x^i}} \partial_{x^j} = \Gamma_{ij}{}^k \partial_{x^k}$ in terms of the *Christoffel symbols*, $\Gamma = {}^\nabla\Gamma := (\Gamma_{ij}{}^k)$. Let $[\cdot, \cdot]$ be the *Lie bracket*. Let $\xi_i \in C^\infty(TM)$ be smooth vector fields on M. The *torsion tensor* is defined by setting

$$\mathcal{T}(\xi_1, \xi_2) := \nabla_{\xi_1} \xi_2 - \nabla_{\xi_2} \xi_1 - [\xi_1, \xi_2] .$$

The torsion $\mathcal{T}$ is tensorial; if $f \in C^\infty(M)$, then $\mathcal{T}(f\xi_1, \xi_2) = \mathcal{T}(\xi_1, f\xi_2) = f\mathcal{T}(\xi_1, \xi_2)$. One has the following useful fact which is a restatement of Lemma 3.5 of Book I; it permits one to normalize the choice of the frame so that only the second derivatives of the Christoffel symbols enter into the computation of the curvature.

Lemma 9.1 Let ∇ be a connection on TM. The following conditions are equivalent and if either condition is satisfied, then ∇ is said to be an *affine connection* or a *torsion-free connection*. The pair (M, ∇) is then said to be an *affine manifold*.

1. $\mathcal{T}(\xi_1, \xi_2) = 0$ for all $\xi_i \in T_P M$ and for all $P \in M$.
2. There exist local coordinates for M centered at any $P \in M$ so that $\Gamma_{ij}{}^k(P) = 0$.

Since $\mathcal{T} = \frac{1}{2}\Gamma_{ij}{}^k(dx^i \wedge dx^j) \otimes \partial_{x^k}$, ∇ is torsion-free if and only if one has the symmetry $\Gamma_{ij}{}^k = \Gamma_{ji}{}^k$. We will assume ∇ is torsion-free henceforth; we will not consider connections with torsion in this volume.

Definition 9.2 Let ∇ be a torsion-free connection. The curvature operator $\mathcal{R}$ (see Section 3.2.3 of Book I) and the Ricci tensor ρ are given, respectively, by setting:

$$\mathcal{R}(\xi_1, \xi_2) := \nabla_{\xi_1}\nabla_{\xi_2} - \nabla_{\xi_2}\nabla_{\xi_1} - \nabla_{[\xi_1,\xi_2]} \quad \text{and} \quad \rho(\xi_1, \xi_2) := \operatorname{Tr}\{\xi_3 \to \mathcal{R}(\xi_3, \xi_1)\xi_2\}.$$

If ∇ is the Levi–Civita connection of a pseudo-Riemannian metric g (see Section 9.1.2), then the scalar curvature τ is given by $\tau = \operatorname{Tr}_g\{\rho\}$. The Ricci tensor need not be symmetric in general; we define the symmetric Ricci tensor ρ_s and the alternating Ricci tensor ρ_a by setting

$$\rho_s(\xi_1, \xi_2) = \tfrac{1}{2}\{\rho(\xi_1, \xi_2) + \rho(\xi_2, \xi_1)\} \quad \text{and} \quad \rho_a(\xi_1, \xi_2) = \tfrac{1}{2}\{\rho(\xi_1, \xi_2) - \rho(\xi_2, \xi_1)\}.$$

Expand $\mathcal{R}(\partial_{x^i}, \partial_{x^j})\partial_{x^a} = R_{ija}{}^b\partial_{x^b}$. By Theorem 3.7 and the discussion in Section 3.2.3 of Book I one has:

$$\begin{aligned}
&\mathcal{R} \text{ is tensorial: } \mathcal{R}(f\xi_1, \xi_2)\xi_3 = \mathcal{R}(\xi_1, f\xi_2)\xi_3 = \mathcal{R}(\xi_1, \xi_2)f\xi_3 = f\mathcal{R}(\xi_1, \xi_2)\xi_3.\\
&\mathcal{R} \text{ is alternating: } \mathcal{R}(\xi_1, \xi_2)\xi_3 = -\mathcal{R}(\xi_2, \xi_1)\xi_3.\\
&\text{The Bianchi identity: } \mathcal{R}(\xi_1, \xi_2)\xi_3 + \mathcal{R}(\xi_2, \xi_3)\xi_1 + \mathcal{R}(\xi_3, \xi_1)\xi_2 = 0.\\
&\text{The curvature is local: } R_{ija}{}^b = \partial_{x^i}\Gamma_{ja}{}^b - \partial_{x^j}\Gamma_{ia}{}^b + \Gamma_{ic}{}^b\Gamma_{ja}{}^c - \Gamma_{jc}{}^b\Gamma_{ia}{}^c.
\end{aligned} \tag{9.1.a}$$

One can covariantly differentiate the curvature operator to define $\nabla\mathcal{R}$:

$$(\nabla_{\xi_1}\mathcal{R})(\xi_2, \xi_3)\xi_4 := \nabla_{\xi_1}\{\mathcal{R}(\xi_2, \xi_3)\xi_4\} - \mathcal{R}(\nabla_{\xi_1}\xi_2, \xi_3)\xi_4 - \mathcal{R}(\xi_2, \nabla_{\xi_1}\xi_3)\xi_4 - \mathcal{R}(\xi_2, \xi_3)\nabla_{\xi_1}\xi_4.$$

Higher-order covariant derivatives of the curvature operator are defined recursively in a similar fashion. One obtains the *second Bianchi identity*:

$$\{\nabla_{\xi_1}\mathcal{R}(\xi_2, \xi_3) + \nabla_{\xi_2}\mathcal{R}(\xi_3, \xi_1) + \nabla_{\xi_3}\mathcal{R}(\xi_1, \xi_2)\}\xi_4 = 0.$$

A curve γ in M is said to be a *geodesic* if it satisfies the equation $\nabla_{\dot\gamma}\dot\gamma = 0$. We may express $\gamma(t) = (x^1(t), \ldots, x^m(t))$ in a system of local coordinates. Then γ is a geodesic if and only if it satisfies the Ordinary Differential Equation (ODE)

$$\ddot{x}^i + \Gamma_{jk}{}^i \dot{x}^j \dot{x}^k = 0 \quad \text{for} \quad 1 \le i \le m\,.$$

By the fundamental theorem of ODEs, given $P \in M$ and $\xi \in T_P M$, there exists a unique geodesic γ defined on some interval $(-\epsilon, \epsilon)$ so $\gamma(0) = P$ and $\dot\gamma(0) = \xi$. We say that (M, ∇) is *complete* if every geodesic is defined for $-\infty < t < \infty$. Fix a point P of M. The exponential map $\exp_P$ is the germ of a diffeomorphism defined on a neighborhood of 0 in $T_P M$ to a neighborhood of P in M which is characterized by the property that $\exp_P(t\xi)$ is the geodesic in M starting at P with initial direction ξ. If $(\xi_1, \ldots, \xi_m)$ is a basis for $T_P M$, then

$$\phi(x^1, \ldots, x^m) := \exp_P(x^1 \xi_1 + \cdots + x^m \xi_m)$$

defines a system of coordinates on M which are called *geodesic coordinates*; they are characterized by the property that straight lines through the origin are geodesics.

9.1.2 PSEUDO-RIEMANNIAN GEOMETRY. Let (M, g) be a pseudo-Riemannian manifold. Here, M is a smooth m-dimensional manifold and g is a non-degenerate symmetric bilinear form on TM of signature (p, q). We refer to the discussion in Chapter 3 of Book I for further details. If $\vec{x} = (x^1, \ldots, x^m)$ is a system of local coordinates on M, expand:

$$g = g_{ij} dx^i \otimes dx^j \quad \text{where} \quad g_{ij} := g(\partial_{x^i}, \partial_{x^j})\,.$$

The pseudo-Riemannian measure is given by $dv_g := |\det(g_{ij})|^{1/2} dx^1 \cdot \cdots \cdot dx^m$; the absolute value is not necessary in the Riemannian setting. The following is a useful technical observation which we will need subsequently. Although it is well-known, we present the proof to keep our treatment self-contained.

Lemma 9.3 Let (M, g) be a pseudo-Riemannian manifold of signature (p, q). There exist smooth complementary subbundles $V_\pm$ of TM so that $TM = V_- \oplus V_+$, so that V_+ is perpendicular to V_-, so that the restriction of g to V_+ is positive definite, and so that the restriction of g to V_- is negative definite.

Proof. Let g_r be an auxiliary Riemannian metric on M. Express $g(\xi_1, \xi_2) = g_r(\Theta\xi_1, \xi_2)$ where Θ is an invertible linear map of the tangent bundle which is self-adjoint with respect to g_r. The bundles $V_\pm$ can then be taken to be the span of the eigenvectors of Θ corresponding to positive/negative eigenvalues of Θ. □

The *Levi–Civita connection* $\nabla = {}^g\nabla$ is the unique torsion-free Riemannian connection on the tangent bundle of M (see Section 3.3.1 of Book I). It is characterized by:

$$\xi_1 g(\xi_2, \xi_3) = g(\nabla_{\xi_1}\xi_2, \xi_3) + g(\xi_2, \nabla_{\xi_1}\xi_3) \quad \text{and} \quad \nabla_{\xi_1}\xi_2 - \nabla_{\xi_2}\xi_1 = [\xi_1, \xi_2]\,.$$

Since ${}^g\nabla$ is torsion-free, $(M, {}^g\nabla)$ is an affine manifold. The associated Ricci tensor ρ is always symmetric in this setting; this need not be true more generally in the affine setting. Introduce the following notation for the first and second derivatives of the metric:

$$g_{ij/k} := \partial_{x^k} g(\partial_{x^i}, \partial_{x^j}) \quad \text{and} \quad g_{ij/k\ell} := \partial_{x^\ell} g_{ij/k}\,.$$

Let g^{ij} be the inverse matrix. Applying the Koszul formula to the coordinate frame yields the *Christoffel identity* (see Equation (3.3.a) of Book I):

$$\Gamma_{ij}{}^k = \tfrac{1}{2} g^{k\ell}\{g_{j\ell/i} + g_{i\ell/j} - g_{ij/\ell}\}\,. \tag{9.1.b}$$

9.1.3 THE RELATION BETWEEN THE LEVI–CIVITA CONNECTION AND THE METRIC.

Although the pseudo-Riemannian metric g defines the Levi–Civita connection ${}^g\nabla$, the Levi–Civita connection does not determine the metric. For example, Equation (9.1.b) shows that ${}^g\nabla = {}^{\lambda g}\nabla$ for any $0 \neq \lambda \in \mathbb{R}$; the metrics g and λg are said to be *homothetic metrics*. More generally, suppose that g and $\hat{g}$ are two pseudo-Riemannian metrics with ${}^g\nabla = {}^{\hat{g}}\nabla$. Define an automorphism Θ of the tangent bundle by requiring that $g(\Theta X, Y) = \hat{g}(X, Y)$. Since $\hat{g}$ is symmetric, Θ is g-self-adjoint. Since $\nabla g = \nabla\hat{g} = 0$, Θ is parallel.

In the Riemannian setting, Θ has constant non-zero eigenvalues λ_i with parallel eigenspaces. Consequently, (M, g) splits locally as a product so that $g = \lambda_1^2 g_1 \oplus \cdots \oplus \lambda_k^2 g_k$ for some $0 \neq \lambda_i \in \mathbb{R}$, $i = 1, \ldots, k$. Therefore, the metric is unique (up to homotheties) for the Levi–Civita connection in the irreducible case.

The higher signature case is more complicated. Any linear map is self-adjoint with respect to some non-degenerate inner product. Consequently, the Jordan normal form of Θ may be quite complicated and Θ may have zero or even complex eigenvalues. The Lorentzian setting is particularly simple as the Jordan normal form of Θ cannot be too complicated. If (M, g) is a Lorentzian manifold which admits no local de Rham product decomposition, results of Martin and Thompson [104] show that any other Lorentzian metric $\hat{g}$ with the same Levi–Civita connection must be (up to homothety) of the form $\hat{g} = g + \lambda\eta\otimes\eta$ where $\lambda \in \mathbb{R}$ and $\eta(X) = g(X, U)$ for some parallel null vector field U.

In addition to the relations of Equation (9.1.a), we have the additional symmetry for the curvature operator of the Levi–Civita connection:

$$\mathcal{R} \text{ is skew-symmetric: } g(\mathcal{R}(\xi_1, \xi_2)\xi_3, \xi_4) + g(\xi_3, \mathcal{R}(\xi_1, \xi_2)\xi_4) = 0\,.$$

We use the metric to lower indices and define

$$R(\xi_1, \xi_2, \xi_3, \xi_4) := g(\mathcal{R}(\xi_1, \xi_2)\xi_3, \xi_4)\,.$$

One may expand $R = R_{ijk\ell}dx^i \otimes dx^j \otimes dx^k \otimes dx^\ell$ for $R_{ijk\ell} := R(\partial_{x^i}, \partial_{x^j}, \partial_{x^k}, \partial_{x^\ell})$. Let $(x^1, \dots, x^m)$ be *geodesic coordinates* centered at a point P of M. In such a coordinate system, the first derivatives of the metric vanish at the center of the coordinate system and we have (see Lemma 3.13 of Book I)

$$R_{ijk\ell}(0) = \tfrac{1}{2}\{\partial_{x^i}\partial_{x^k}g_{j\ell} + \partial_{x^j}\partial_{x^\ell}g_{ik} - \partial_{x^i}\partial_{x^\ell}g_{jk} - \partial_{x^j}\partial_{x^k}g_{i\ell}\}(0)\,.$$

Conversely, one may show (see Lemma 3.8 of Book I) that

$$g_{ij/k}(0) = 0 \quad \text{and} \quad g_{ij/k\ell}(0) = -\tfrac{1}{3}(R_{ikj\ell}(0) + R_{i\ell jk}(0))\,.$$

Clearly, all the components of the covariant derivatives of the curvature can be expressed in terms of the derivatives of the metric. Less trivially, in geodesic coordinates, all the derivatives of the metric at the origin can be expressed in terms of the covariant derivatives of the curvature (see, for example, Atiyah, Bott and Patodi [8]).

9.1.4 VARIOUS NOTIONS OF HOMOGENEITY.

Definition 9.4 A pseudo-Riemannian manifold $\mathcal{M} := (M, g)$ is said to be *locally homogeneous* if for any pair of points P and Q in M, there exists the germ of an isometry $\phi_{P,Q}$ taking P to Q. This implies that $\phi_{P,Q}^*\nabla^i\mathcal{R}_Q = \nabla^i\mathcal{R}_P$ for all i, i.e., the curvature operator and all its covariant derivatives "look the same at every point". We say that $\mathcal{M}$ is *k-curvature homogeneous* if given any pair of points P and Q in M, there exists a linear isometry $\Phi_{P,Q}$ taking (T_PM, g_P) to (T_QM, g_Q) so that $\Phi_{P,Q}^*\nabla^i\mathcal{R}_Q = \nabla^i\mathcal{R}_P$ for $0 \le i \le k$. Since $\Phi_{P,Q}$ is a linear isometry, we can pass from the curvature operator to the corresponding curvature tensor and require equivalently that $\Phi_{P,Q}^*\nabla^i R_Q = \nabla^i R_P$ for $0 \le i \le k$. This passage is, of course, not possible in the affine setting. In the homothety curvature setting it imposes some rescaling requirements as we will see subsequently in Lemma 9.21.

It is convenient to pass to the purely algebraic context.

Definition 9.5 A *k-curvature model* is a collection $\mathfrak{M}_k := (V, \langle\cdot,\cdot\rangle, A^0, \dots, A^k)$ where $\langle\cdot,\cdot\rangle$ is a non-degenerate symmetric bilinear form on an m-dimensional real vector space V and where $A^i \in \otimes^{4+i}(V^*)$ is to model the ith covariant derivative of the curvature tensor $\nabla^i R$. One sometimes imposes the additional relations generated by the usual $\mathbb{Z}_2$-symmetries and the Bianchi identities; we will not do that as it is a bit of extra fuss that is only needed if one wants to prove geometric realization theorems. Two k-curvature models are said to be *isomorphic k-curvature models* if there exists a linear isomorphism Φ from V^1 to V^2 so that $\Phi^*\langle\cdot,\cdot\rangle^2 = \langle\cdot,\cdot\rangle^1$ and so that $\Phi^*A^{i,2} = A^{i,1}$ for $0 \le i \le k$. We say that $\mathfrak{M}_k$ is a k-curvature model for a pseudo-Riemannian manifold $\mathcal{M} = (M, g)$ if $(T_PM, g_P, R_P, \dots, \nabla^k R_P)$ is isomorphic to $\mathfrak{M}_k$ for any point P of M. Clearly, $\mathcal{M}$ is *k-curvature homogeneous* if and only if $\mathcal{M}$ admits a k-curvature model as the model could be taken to be $(T_PM, g_P, R_P, \dots, \nabla^k R_P)$ for any point P of M.

Definition 9.6 A pseudo-Riemannian manifold $\mathcal{M}$ is said to be *locally symmetric* if one has that $\nabla\mathcal{R} = 0$. Let $\nabla^2\mathcal{R}(X_1, X_2; X_3, X_4)$ be the second covariant derivative of the curvature operator. Commuting covariant derivatives introduces curvature. If $\mathcal{M}$ is locally symmetric, then $\nabla^2\mathcal{R} = 0$ so that for all X_i one has

$$\begin{aligned} 0 &= \nabla^2\mathcal{R}(X_1, X_2; X_3, X_4) - \nabla^2\mathcal{R}(X_1, X_2; X_4, X_3) \\ &= \mathcal{R}(\mathcal{R}(X_3, X_4)X_1, X_2) + \mathcal{R}(X_1, \mathcal{R}(X_3, X_4)X_2) . \end{aligned} \tag{9.1.c}$$

The vanishing on the second line of this display is a condition on the 0-curvature model of a symmetric space. More generally, $\mathcal{M}$ is said to be *semi-symmetric* if this condition holds. This is a purely algebraic condition which is defined on the 0-model. Note that there are semi-symmetric spaces which are not symmetric.

9.1.5 CURVATURE HOMOGENEOUS SPACES WHICH ARE NOT LOCALLY HOMOGENEOUS.

Clearly, any 0-curvature homogeneous surface has constant Gauss curvature. This implies that it is locally symmetric. Hence, we always assume $\dim(M) \geq 3$. An specific feature of dimension 3 is that, since the Weyl tensor (see Definition 11.1) vanishes, the curvature tensor is determined by the Ricci tensor. Hence, a 3-dimensional pseudo-Riemannian manifold is 0-curvature homogeneous if and only if the Ricci operator $g(\text{Ric}(x), y) = \rho(x, y)$ has constant eigenvalues. The first non-homogeneous example was given by Sekigawa [124].

Example 9.7 Let $\mathcal{M} = (\mathbb{R}^2 \times \mathbb{R}, g_{\mathbb{R}^2} + \psi g_{\mathbb{R}})$ where $\psi(x^1, x^2, x^3) = e^{2x^1 \cos(x^3) - 2x^2 \sin(x^3)}$. The Ricci operator of this warped product has constant eigenvalues $\{-1, -1, 0\}$. Consequently, $\mathcal{M}$ is 0-curvature homogeneous. Since $\|\nabla\rho\|^2 = 2e^{-2x^1 \cos(x^3) + 2x^2 \sin(x^3)}$, $\mathcal{M}$ is not 1-curvature homogeneous therefore not locally homogeneous. The 0-curvature model $\mathfrak{M}_0$ is that of the symmetric space $\mathbb{H}^2 \times \mathbb{R}$.

Example 9.7 presents a curvature homogeneous manifold which is not homogeneous but which has the same 0-curvature model as that of a homogeneous space. The following example of Lastaria [99] exhibits a continuous family of non-isometric curvature homogeneous manifolds all which have the same 0-curvature model.

Example 9.8 Let $\{e_1, e_2, e_3\}$ be the standard basis for the Lie algebra $\mathfrak{so}(3)$. One then has that $[e_1, e_2] = e_3$, $[e_2, e_3] = e_1$, and $[e_3, e_1] = e_2$. Let ω_i be the dual basis. For $\alpha \geq 1$, let

$$g_\alpha = \frac{\alpha^2}{1+\alpha^2}\omega_1 \otimes \omega_1 + \frac{1}{1+\alpha^2}\omega_2 \otimes \omega_2 + \omega_3 \otimes \omega_3$$

be a smooth 1-parameter family of left-invariant metrics on $SO(3)$. The Ricci eigenvalues of these metrics are $\{0, 0, 2\}$. Consequently, all these metrics have the same 0-curvature model. However, since $\|\nabla\rho\|^2 = 8\frac{1+\alpha^4}{\alpha^2}$, the metrics in this family are not locally isometric.

Author's Biography

AHMED ELGAMMAL

Dr. Elgammal is an associate professor at the Department of Computer Science, Rutgers, the State University of New Jersey since Fall 2002. Dr. Elgammal is also a member of the Center for Computational Biomedicine Imaging and Modeling (CBIM). His primary research interest is computer vision and machine learning. His research focus includes human activity recognition, human motion analysis, tracking, human identification, and statistical methods for computer vision. Dr. Elgammal received the National Science Foundation CAREER Award in 2006. He has been the principal investigator and co-principal investigator of several research projects in the areas of human motion analysis, gait analysis, tracking, facial expression analysis and scene modeling; funded by NSF and ONR. Dr. Elgammal is a member of the review committee/board in several of the top conferences and journals in the computer vision field and is a senior IEEE member. Dr. Elgammal received his Ph.D. in 2002 and M.Sc. degree in 2000 in computer science from the University of Maryland, College Park. Dr. Elgammal received M.Sc. and B.E. degrees in computer science and automatic control from Alexandria University in 1996 and 1993, respectively.

corresponds to a symmetric space. In that setting, $\nabla^k R = 0$ for all $k \geq 1$ so all the information is encoded in the 0-curvature model. If the symmetric space is irreducible then the corresponding model is Einstein and one has the following result of Boeckx, Kowalski and Vanhecke [14].

Theorem 9.13 *Let (M, g) be a Riemannian space with the same curvature tensor as an irreducible symmetric space. Then (M, g) is locally symmetric and locally isometric to its model space.*

Recall that $\mathcal{M}$ is called *semi-symmetric* if Equation (9.1.c) holds. The structure theorems of Szabó for semi-symmetric spaces [129] and work of Boeckx, Kowalski and Vanhecke [14] yield the following result.

Theorem 9.14 *Let (M, g) be a semi-symmetric curvature homogeneous Riemannian manifold. There exists a symmetric space (M_s, g_s) and there exist locally irreducible Riemannian spaces (F_i, g_i) which are foliated by totally geodesic Euclidean leaves of codimension two and have constant scalar curvature so that (M, g) is locally isometric to the Riemannian product*

$$(M_s, g_s) \times (F_1, g_1) \times \cdots \times (F_r, g_r) .$$

It follows from Theorem 9.14 that for semi-symmetric curvature homogeneous Riemannian manifold which is locally irreducible and not locally homogeneous, the curvature tensor is modeled in a cylinder $S^2 \times \mathbb{R}^m$ or $\mathbb{H}^2 \times \mathbb{R}^m$. Further, note that there is an infinite-dimensional family of locally non-isometric Riemannian spaces as above. We will refer to work of Boeckx, Kowalski and Vanhecke [14] and Gilkey [73] for more information on curvature homogeneous manifolds.

Definition 9.15 We say that an affine manifold $\mathcal{M} = (M, \nabla)$ is *locally affine homogeneous* if for any pair of points P and Q in M, there exists the germ of a diffeomorphism $\phi_{P,Q}$ taking P to Q so that $\phi^*_{P,Q}\nabla = \nabla$. We say that $\mathcal{M}$ is *affine k-curvature homogeneous* if given any pair of points P and Q in M, there is a linear isomorphism $\Phi_{P,Q}$ from $T_P M$ to $T_Q M$ so that $\Phi^*_{P,Q}\nabla^i \mathcal{R}_Q = \nabla^i \mathcal{R}_P$ for $0 \leq i \leq k$.

Let $\mathcal{M} := (M, g)$ be a pseudo-Riemannian manifold. We can define an associated affine manifold (M, ∇) by taking ∇ to be the Levi–Civita connection of g. Clearly, if (M, g) is locally homogeneous, then (M, ∇) is locally affine homogeneous. On the other hand, the converse need not be true since the group of affine transformations is larger than the group of isometries. We suppose that (M, g) is locally affine homogeneous and let $\phi_{P,Q}$ be the germ of a diffeomorphism taking P to Q so that $\phi^*_{P,Q}\nabla = \nabla$. Let $\hat{g}$ be the pseudo-Riemannian metric $\hat{g} = \phi^*_{P,Q} g$ with Levi–Civita connection $\phi^*_{P,Q}\nabla$. Since $\phi^*_{P,Q}\nabla = \nabla$ both metrics g and $\hat{g}$ share the same Levi–Civita connection. The discussion of Section 9.1.3 shows that g and $\hat{g}$ are homothetical in the

Riemannian irreducible setting (and, moreover, in the indecomposable Lorentzian situation). This shows that if (M, g) is indecomposable, then (M, ∇) is locally affine homogeneous if and only if (M, g) is locally homothety homogeneous (see Definition 9.16) both in the Riemannian and Lorentzian signatures. We refer to the discussion in Kowalski, Vlášek and Opozda [93] for further details.

9.1.6 MANIFOLDS WHICH ARE HOMOTHETY HOMOGENEOUS.

There is a slightly weaker notion due to Kowalski and Vanžurová [91, 92] that will play an important role in our subsequent discussion; it lies between affine k-curvature homogeneity and k-curvature homogeneity.

Definition 9.16 We say that a pseudo-Riemannian manifold is *locally homothety homogeneous* if for any pair of points P and Q in M, there exists the germ of a diffeomorphism $\phi_{P,Q}$ taking P to Q so that $\phi_{P,Q}$ is a *homothety*; this means that $\phi_{P,Q}^* g = \lambda_{P,Q}^2 g$ for some $\lambda_{P,Q} \neq 0$.

We postpone for the moment a precise definition of k-homothety curvature homogeneity but simply note that it is a weaker notion than k-curvature homogeneity and a stronger notion that affine k-curvature homogeneity, i.e.,

$$\begin{array}{ccccc}
\text{homog.} & \Rightarrow & \text{homothety homog.} & \Rightarrow & \text{affine homog.} \\
\Downarrow & & \Downarrow & & \Downarrow \\
k\text{-curv. homog.} & \Rightarrow & k\text{-homothety curv. homog.} & \Rightarrow & \text{affine } k\text{-curv. homog.}
\end{array}$$

It is relatively easy to write down examples in the indefinite setting; we will present two examples and refer to Gilkey [73] for others. We first begin with an example that is 0-curvature homogeneous (and, consequently, affine 0-curvature homogeneous and 0-homothety curvature homogeneous as these are weaker properties) but which is not affine 1-curvature homogeneous (and, consequently, neither 1-curvature homogeneous nor 1-homothety curvature homogeneous as these are stronger properties). The following example is due to Gilkey and Nikčević [74].

Example 9.17 Let $(u^1, u^2, t^1, t^2, v^1, v^2)$ be coordinates on $\mathbb{R}^6$. Let $f_1(u^1)$ and $f_2(u^2)$ be smooth functions of 1-variable. Let $\mathcal{M}_{f_1,f_2} := (\mathbb{R}^6, g_{f_1,f_2})$ be the Ricci flat pseudo-Riemannian manifold of signature $(4, 2)$ defined by:

$$\begin{aligned}
&g_{f_1,f_2}(\partial_{u^i}, \partial_{u^i}) = -2f_1(u^1) - 2f_2(u^2) - 2u^1t^1 - 2u^2t^2, \\
&g_{f_1,f_2}(\partial_{u^i}, \partial_{v^i}) = 1, \quad g_{f_1,f_2}(\partial_{t^i}, \partial_{t^i}) = -1\,.
\end{aligned}$$

The non-zero entries in R and ∇R are, up to the usual $\mathbb{Z}_2$-symmetries:

$$\begin{aligned}
&R(\partial_{u^1}, \partial_{u^2}, \partial_{u^2}, \partial_{u^1}) = f_1''(u^1) + f_2''(u^2) + (u^1)^2 + (u^2)^2, \quad R(\partial_{u^1}, \partial_{u^2}, \partial_{u^2}, \partial_{t^1}) = 1, \\
&\nabla R(\partial_{u^1}, \partial_{u^2}, \partial_{u^2}, \partial_{u^1}; \partial_{u^i}) = f_i'''(u^i) + 4u^i\,.
\end{aligned}$$

One can show that $\mathcal{M}_{f_1,f_2}$ is 0-curvature homogeneous. One has that $\mathcal{M}_{f_1,f_2}$ is 1-affine curvature homogeneous if and only if $f_1^{(4)} = f_2^{(4)} = -4$ and in fact $\mathcal{M}_{f_1,f_2}$ is homogeneous in this setting. Consequently, 0-curvature homogeneity does not imply 1-affine curvature homogeneity. We refer to Gilkey [73] §2.7 for further details where we take $s = 2$ in the notation of that paper.

More generally, there are pseudo-Riemannian manifolds which are k-curvature homogeneous but not $(k+1)$-curvature homogeneous for any $k \geq 0$. In Section 9.5, we will discuss *generalized plane wave manifolds*. We say that $\mathcal{M}$ is *VSI* if all the scalar Weyl invariants vanish. We will show that any generalized plane wave manifold is geodesically complete and VSI. We will show in Lemma 9.47 that the manifolds given below in Example 9.18 are generalized plane wave manifolds. Thus, they are geodesically complete and VSI. We will also use a manifold in this family in Section 10.1 (see Lemma 10.2) to construct an example of a homogeneous manifold with a non-trivial homothety character. These are examples which are k-curvature homogeneous (and, consequently, affine k-curvature homogeneous and k-homothety curvature homogeneous) but which are not affine $(k+1)$-curvature homogeneous (and, consequently, neither $(k+1)$-curvature homogeneous nor $(k+1)$-homothety curvature homogeneous). The following example is from Gilkey and Nikčević [75].

Example 9.18 For $\ell \geq -1$, let $(x, y, z^0, \dots, z^\ell, \tilde{x}, \tilde{y}, \tilde{z}^0, \dots, \tilde{z}^\ell)$ be coordinates on $\mathbb{R}^{6+2\ell}$ (if $\ell = -1$, then the z variables are not present). Let $f(y)$ be a smooth function of one variable with $f^{(\ell+3)} > 0$ and $f^{(\ell+4)} > 0$. Let $\mathcal{M}_f := (\mathbb{R}^{6+2\ell}, g_f)$ be the pseudo-Riemannian manifold of signature $(\ell+3, \ell+3)$ defined by:

$$\begin{aligned} &g_f(\partial_x, \partial_x) := -2\{f(y) + yz^0 + \dots + y^{\ell+1}z^\ell\}, \\ &g_f(\partial_x, \partial_{\tilde{x}}) := 1, \ g_f(\partial_y, \partial_{\tilde{y}}) := 1, \ g_f(\partial_{z^i}, \partial_{\tilde{z}^i}) := 1\,. \end{aligned}$$

Then $\mathcal{M}_f$ is $(\ell+2)$-curvature homogeneous. Furthermore, $\mathcal{M}_f$ is affine $(\ell+3)$-curvature homogeneous if and only if $f^{(\ell+3)} = ae^{by}$ for some $a > 0$ and $b > 0$ and $\mathcal{M}_f$ is in fact homogeneous in this setting. Thus, generically, $\mathcal{M}_f$ is a $(\ell+2)$-curvature homogeneous manifold which is not $(\ell+3)$-curvature homogeneous. We refer as well to the discussion in Gilkey [73] §2.9 for further details.

Remark 9.19 We will define the *modified Riemannian extension* subsequently in Chapter 11 (see Definition 11.39) and simply note the following in passing. Let ϕ be the symmetric tensor field whose only non-zero component is $\phi(\partial_x, \partial_x) = -2\{f(y) + yz^0 + \dots + y^{\ell+1}z^\ell\}$ and let ∇ be the usual flat connection on $\mathbb{R}^{\ell+1}$. Identify $\mathbb{R}^{2\ell+2}$ with the cotangent bundle of $\mathbb{R}^{\ell+1}$. The metric g_f is then the modified Riemannian extension $g_{\nabla,\phi}$.

We postpone until Section 10.2 a more detailed discussion of homothety homogeneous manifolds and content ourselves for the moment with the following example which motivates the discussion of Chapter 10.

Theorem 9.20 *Let $\mathcal{N} = (N, g_N)$ be a pseudo-Riemannian manifold which has dimension $m-1$ where $m \geq 3$. Let $\mathcal{M}_t := (M, g_t)$ where $M = \mathbb{R} \times N$ and $g_t := e^{tx}(dx^2 + g_N)$.*

1. $\tau_{\mathcal{M}_t} = e^{-tx}\{\tau_{\mathcal{N}} - \frac{(m-1)(m-2)}{4}t^2\}$.
2. *If $\mathcal{N}$ is homogeneous, then $\mathcal{M}_t$ is homothety homogeneous for any t and, for generic t, $\mathcal{M}_t$ is not homogeneous.*

We note that if $\mathcal{N}$ is homogeneous, then (M, ∇^t) is affine homogeneous for all t, while (M, g_t) is not locally homogeneous for generic metrics g_t.

Proof. Although results of Alekseevsky et al. [4] would permit one to compute τ_M, it is instructive to establish Assertion 1 via a direct computation. We examine the curvature tensor. Fix a point $P \in N$. Choose local coordinates $\vec{y} = (y^1, \dots, y^{m-1})$ centered at P. Let indices u, v, w range from 0 to $m-1$ and index the coordinate frame $(\partial_x, \partial_{y^1}, \dots, \partial_{y^{m-1}})$ and let indices i, j, k range from 1 to $m-1$ and index the coordinate frame $(\partial_{y^1}, \dots, \partial_{y^{m-1}})$. Let Γ be the Christoffel symbols of g_N and $\tilde{\Gamma}$ be the Christoffel symbols of g_M. Let δ_{ij} be the Kronecker index. We compute:

$$\begin{array}{lllll} \tilde{g}_{00} = e^{tx}, & \tilde{g}_{0i} = 0, & \tilde{g}_{ij} = e^{tx} g_{ij}, & \tilde{\Gamma}_{00}{}^0 = \frac{1}{2}t & \tilde{\Gamma}_{00}{}^i = 0, \\ \tilde{\Gamma}_{0i}{}^0 = 0, & \tilde{\Gamma}_{0i}{}^k = \frac{1}{2}t\delta_{ik}, & \tilde{\Gamma}_{ij}{}^0 = -\frac{1}{2}tg_{ij}, & \tilde{\Gamma}_{ij}{}^k = \Gamma_{ij}{}^k, & \tilde{\nabla}_{\partial_x}\partial_x = \frac{1}{2}t\partial_x, \\ \tilde{\nabla}_{\partial_x}\partial_{y^i} = \frac{1}{2}t\partial_{y^i}, & \tilde{\nabla}_{\partial_{y^i}}\partial_x = \frac{1}{2}t\partial_{y^i}, & \tilde{\nabla}_{\partial_{y^i}}\partial_{y^j} = \Gamma_{ij}{}^k\partial_{y^k} - \frac{1}{2}tg_{ij}\partial_x . & & \end{array}$$

We choose the coordinate system so the first derivatives of g_{ij} vanish at P and, consequently, $\Gamma(P) = 0$. Consequently, the curvature operator at P is given by:

$$\tilde{\mathcal{R}}(\partial_x, \partial_{y^i})\partial_x = \{\tilde{\nabla}_{\partial_x}\tilde{\nabla}_{\partial_{y^i}} - \tilde{\nabla}_{\partial_{y^i}}\tilde{\nabla}_{\partial_x}\}\partial_x = \tfrac{1}{2}t\tilde{\nabla}_{\partial_x}\partial_{y^i} - \tfrac{1}{2}t\tilde{\nabla}_{\partial_{y^i}}\partial_x = 0,$$

$$\begin{aligned} \tilde{\mathcal{R}}(\partial_x, \partial_{y^i})\partial_{y^j} &= \{\tilde{\nabla}_{\partial_x}\tilde{\nabla}_{\partial_{y^i}} - \tilde{\nabla}_{\partial_{y^i}}\tilde{\nabla}_{\partial_x}\}\partial_{y^j} \\ &= \tilde{\nabla}_{\partial_x}\{\Gamma_{ij}{}^k\partial_{y^k} - \tfrac{1}{2}tg_{ij}\partial_x\} - \tfrac{1}{2}t\tilde{\nabla}_{\partial_{y^i}}\partial_{y^j} \\ &= \tfrac{1}{2}t\{\Gamma_{ij}{}^k\partial_{y^k} - \tfrac{1}{2}tg_{ij}\partial_x\} - \tfrac{1}{2}t\{\Gamma_{ij}{}^k\partial_{y^k} - \tfrac{1}{2}tg_{ij}\partial_x\} = 0, \end{aligned}$$

$$\tilde{\mathcal{R}}(\partial_{y^i}, \partial_{y^j})\partial_x = \{\tilde{\nabla}_{\partial_{y^i}}\tilde{\nabla}_{\partial_{y^j}} - \tilde{\nabla}_{\partial_{y^j}}\tilde{\nabla}_{\partial_{y^i}}\}\partial_x = \tfrac{1}{2}t\tilde{\nabla}_{\partial_{y^i}}\partial_{y^j} - \tfrac{1}{2}t\tilde{\nabla}_{\partial_{y^j}}\partial_{y^i} = 0,$$

$$\begin{aligned} \tilde{\mathcal{R}}(\partial_{y^i}, \partial_{y^j})\partial_{y^k} &= \{\tilde{\nabla}_{\partial_{y^i}}\tilde{\nabla}_{\partial_{y^j}} - \tilde{\nabla}_{\partial_{y^j}}\tilde{\nabla}_{\partial_{y^i}}\}\partial_{y^k} \\ &= \tilde{\nabla}_{\partial_{y^i}}(\Gamma_{jk}{}^\ell\partial_{y^\ell} - \tfrac{1}{2}tg_{jk}\partial_x) - \tilde{\nabla}_{\partial_{y^j}}(\Gamma_{ik}{}^\ell\partial_{y^\ell} - \tfrac{1}{2}tg_{ik}\partial_x) \\ &= R_{ijk}{}^\ell\partial_{y^\ell} - \tfrac{1}{4}t^2g_{jk}\partial_{y^i} + \tfrac{1}{4}t^2g_{ik}\partial_{y^j} . \end{aligned}$$

We can now express τ_M in terms of τ_N to complete the proof of Assertion 1.

Suppose $\mathcal{N}$ is homogeneous. We extend the action of the isometry group of N to an isometric action on M preserving the slices $\{x\} \times N$. The shift $(x, \xi) \to (x + c, \xi)$ is a homothety and, therefore, $\mathcal{M}_t$ is homothety homogeneous. If $\tau_{\mathcal{N}} - \frac{1}{4}(m-1)(m-2) \neq 0$, then $\tau_{\mathcal{M}_t}$ is not constant and, consequently, $\mathcal{M}_t$ is homothety homogeneous but not homogeneous. □

Recall that an *inner product space* is a pair $(V, \langle\cdot,\cdot\rangle)$ where V is a finite-dimensional vector space and where $\langle\cdot,\cdot\rangle$ is a non-degenerate symmetric bilinear form of signature (p, q) on V. Let $(V_i, \langle\cdot,\cdot\rangle^i)$ be inner product spaces. A linear map Φ from V_1 to V_2 is said to be an isometry if $\Phi^*\langle\cdot,\cdot\rangle^2 = \langle\cdot,\cdot\rangle^1$ and a *homothety* if $\Phi^*\langle\cdot,\cdot\rangle^2 = \lambda^2\langle\cdot,\cdot\rangle^1$ for some $\lambda \neq 0$. The following result reflects the difference between the curvature tensor and the curvature operator; a homothety will preserve the curvature operator but rescale the curvature tensor.

Lemma 9.21 The following conditions are equivalent and if any is satisfied, then a pseudo-Riemannian manifold $\mathcal{M} = (M, g)$ will be said to be k-*homothety curvature homogeneous*.

1. Given any two points $P, Q \in M$, there is a linear homothety $\Phi = \Phi_{P,Q}$ from $(T_P M, g_P)$ to $(T_Q M, g_Q)$ so that if $0 \leq \ell \leq k$, then $\Phi^*(\nabla^\ell \mathcal{R}_Q) = \nabla^\ell \mathcal{R}_P$.
2. Given any two points $P, Q \in M$, there is a linear isometry $\phi = \phi_{P,Q}$ from $T_P M$ to $T_Q M$ and $0 \neq \lambda = \lambda_{P,Q} \in \mathbb{R}$ so that if $0 \leq \ell \leq k$, then $\phi^*(\nabla^\ell R_Q) = \lambda^{-\ell-2}\nabla^\ell R_P$.

We note that this agrees with Proposition 0.1 of Kowalski and Vanžurová [92]. If we can take $\lambda_{P,Q} = 1$ for all P and Q, then $\mathcal{M}$ is k-curvature homogeneous. But as we will see in Theorem 10.27, there are examples which are 2-homothety curvature homogeneous which are not 2-curvature homogeneous. Thus, we cannot take $\lambda_{P,Q} = 1$ for all $P, Q \in M$.

Proof. Assume that Assertion 1 of Lemma 9.21 holds. This means that given any two points P and Q in M, there exists a linear homothety $\Phi : T_P M \to T_Q M$ so that if $0 \leq \ell \leq k$, then $g_Q(\Phi\xi_1, \Phi\xi_2) = \lambda^2 g_P(\xi_1, \xi_2)$ and

$$\Phi\left\{\nabla^\ell \mathcal{R}_P(\xi_1, \xi_2; \xi_5, \ldots, \xi_{\ell+4})\xi_3\right\} = \nabla^\ell \mathcal{R}_Q(\Phi\xi_1, \Phi\xi_2; \Phi\xi_5, \ldots, \Phi\xi_{\ell+4})\Phi\xi_3 .$$

Taking the inner product with $\Phi\xi_4$ permits us to rewrite this identity in the form:

$$\begin{aligned}
&\lambda^2\nabla^\ell R_P(\xi_1, \xi_2, \xi_3, \xi_4; \xi_5, \ldots, \xi_{\ell+4}) \\
&\quad = \lambda^2 g_P\left(\nabla^\ell \mathcal{R}_P(\xi_1, \xi_2; \xi_5, \ldots, \xi_{\ell+4})\xi_3, \xi_4\right) \\
&\quad = g_Q(\Phi\nabla^\ell \mathcal{R}_P(\xi_1, \xi_2; \xi_5, \ldots, \xi_{\ell+4})\xi_3, \Phi\xi_4) \\
&\quad = g_Q(\nabla^\ell \mathcal{R}_Q(\Phi\xi_1, \Phi\xi_2; \Phi\xi_5, \ldots, \Phi\xi_{\ell+4})\Phi\xi_3, \Phi\xi_4) \\
&\quad = \nabla^\ell R_Q(\Phi\xi_1, \Phi\xi_2, \Phi\xi_3, \Phi\xi_4; \Phi\xi_5, \ldots, \Phi\xi_{\ell+4}) .
\end{aligned}$$

We set $\phi := \lambda^{-1}\Phi$. We can rewrite these equations in the form:

$$\begin{aligned}
&g_Q(\phi\xi_1, \phi\xi_2) = \lambda^{-2} g_Q(\Phi\xi_1, \Phi\xi_2) = g_P(\xi_1, \xi_2),\\
&\lambda^2 \nabla^\ell R_P(\xi_1, \xi_2, \xi_3, \xi_4; \xi_5, \dots, \xi_{\ell+4})\\
&\qquad = \nabla^\ell R_Q(\Phi\xi_1, \Phi\xi_2, \Phi\xi_3, \Phi\xi_4; \Phi\xi_5, \dots, \Phi\xi_{\ell+4})\\
&\qquad = \lambda^{\ell+4} \nabla^\ell R_Q(\phi\xi_1, \phi\xi_2, \phi\xi_3, \phi\xi_4; \phi\xi_5, \dots, \phi\xi_{\ell+4})\\
&\qquad = \lambda^{\ell+4} \phi^*(\nabla^\ell R_Q)(\xi_1, \xi_2, \xi_3, \xi_4; \xi_5, \dots, \xi_{\ell+4}) .
\end{aligned}$$

This shows ϕ is an isometry from $T_P M$ to $T_Q M$ so $\phi^*(\nabla^\ell R_Q) = \lambda^{-2-\ell}\nabla^\ell R_P$. Consequently, Assertion 1 $\Rightarrow$ Assertion 2. The proof of the remaining implication is similar and will be omitted. □

Remark 9.22 Modeling at the algebraic level Assertion 2 of Lemma 9.21, we say that two k-curvature models are *k-homothety isomorphic* if there exists $\lambda \neq 0$ and a linear isomorphism ϕ so that $\phi^*\langle\cdot,\cdot\rangle^2 = \langle\cdot,\cdot\rangle^1$ and so that $\phi^* A^{i,1} = \lambda^{-i-2} A_{i,2}$. We say that $\mathfrak{M}_k$ is a k-homothety curvature model for $\mathcal{M}$ if $\mathfrak{M}_k$ is k-homothety isomorphic to $(T_P M, g_P, R_P, \dots, \nabla^k R_P)$ for any point P of M. It is now evident that $\mathcal{M}$ is k-homothety curvature homogeneous if and only if $\mathcal{M}$ admits a k-homothety curvature model.

9.2 INVARIANCE THEORY IN THE RIEMANNIAN SETTING

In Section 9.2.1, we discuss the first and second Theorems of Invariants which are due to Weyl [134]. In Section 9.2.2, we apply these results to discuss scalar invariants in pseudo-Riemannian geometry. In Section 9.2.3 (resp. Section 9.2.4), we discuss universal scalar-valued (resp. symmetric 2-tensor-valued) curvature identities; these results will play a central role in the analysis of Section 9.3.1 when we discuss the Euler–Lagrange equations of the Chern–Gauss–Bonnet integrand. In Section 9.2.5, we establish the results stated in Sections 9.2.3 and 9.2.4.

9.2.1 WEYL'S THEOREMS OF INVARIANTS. Let $\mathcal{O} := \mathcal{O}(V, \langle\cdot,\cdot\rangle)$ be the *orthogonal group* of an inner product space $(V, \langle\cdot,\cdot\rangle)$. A multilinear map ψ from $\times^k V$ to $\mathbb{R}$ is said to be a *linear orthogonal invariant* of degree k if $\Theta^*\psi = \psi$ for all $\Theta \in \mathcal{O}$. If we take $\Theta = -\text{Id}$, then $\Theta^*\psi = (-1)^k\psi$ so orthogonal invariants do not exist if k is odd. We refer to Weyl [134, Thm 2.9.A p. 53, Theorem 2.17.A p. 75] for the proof of the following result.

Hermann Weyl (1885–1955)

Theorem 9.23 (Weyl). *Let $(V, \langle\cdot,\cdot\rangle)$ be an inner product space of dimension m.*

1. **First Theorem of Invariants:** *Let k be even and let $\pi \in \text{Perm}(k)$ be a permutation of the integers from 1 to k. We define an orthogonal invariant $\psi_\pi \in \otimes^k V^*$ by setting*

$$\psi_\pi(v^1, \dots, v^{2\ell}) := \langle v^{\pi(1)}, v^{\pi(2)}\rangle \cdot \dots \cdot \langle v^{\pi(k-1)}, v^{\pi(k)}\rangle .$$

   Then the space of linear orthogonal invariants of degree k is spanned by the maps ψ_π.

2. **Second Theorem of Invariants:** *Every relation among scalar products is an algebraic consequence of the relations*

$$\det\begin{pmatrix} \langle v^0, w^0\rangle & \langle v^0, w^1\rangle & \dots & \langle v^0, w^m\rangle \\ \langle v^1, w^0\rangle & \langle v^1, w^1\rangle & \dots & \langle v^1, w^m\rangle \\ \dots & \dots & \dots & \dots \\ \langle v^m, w^0\rangle & \langle v^m, w^1\rangle & \dots & \langle v^m, w^m\rangle \end{pmatrix} = 0 .$$

9.2.2 SCALAR INVARIANTS OF RIEMANNIAN MANIFOLDS. The following result, which in the Riemannian setting seems to be first due to Atiyah, Bott and Patodi [8], is then a direct consequence of Assertion 1 of Theorem 9.23; the extension to the pseudo-Riemannian setting is not difficult (see, for example, Brozos-Vázquez, Gilkey and Nikčević [26]).

M. F. Atiyah (1929–)

V. K. Patodi (1945–1976)

R. Bott (1923–2005)

Lemma 9.24 All scalar invariants in the category of m-dimensional pseudo-Riemannian manifolds which are given by a local formula in the derivatives of the metric arise by contracting indices in pairs in monomial expressions in the covariant derivatives of the curvature tensor.

There is a natural order which is defined on the space of invariants which plays a crucial role. We say that $R_{ijk}{}^{\ell}$ has order 2 and we increase the order by 1 for each explicit covariant derivative; equivalently, we say the order of $g_{ij/k}$ is 1 and we increase the order by 1 for each explicit partial derivative. Let $\mathcal{I}_{m,n}$ be the space of scalar invariant local formulas which are homogeneous of total order n in the derivatives of the metric and which are defined in the category of all Riemannian manifolds of dimension m; we suppress for the moment the slight additional complexities involved in passing to the pseudo-Riemannian setting. Let $R_{ijk\ell}$ be the components of the curvature tensor relative to a local orthonormal frame $\{e_1, \ldots, e_m\}$ for TM. The scalar curvature $\tau_m := R_{ijji}$ belongs to $\mathcal{I}_{m,2}$. Let ρ be the Ricci tensor and let R be the full curvature tensor. The following result then follows from Lemma 9.24 after a bit of work to eliminate redundancies (see Gilkey [72] for details).

Lemma 9.25

1. $\mathcal{I}_{m,0} = \text{span}\{1\}$.
2. $\mathcal{I}_{m,2} = \text{span}\{\tau := R_{ijji}\}$.
3. $\mathcal{I}_{m,4} = \text{span}\{R_{ijji;kk}, \tau^2 := R_{ijji}R_{k\ell\ell k}, \|\rho\|^2 := R_{ijjk}R_{i\ell\ell k}, \|R\|^2 := R_{ijk\ell}R_{ijk\ell}\}$.
4. $\mathcal{I}_{m,6} = \text{span}\{R_{ijji;kk\ell\ell}, R_{ijji;k}R_{\ell nn\ell;k}, R_{aija;k}R_{bijb;k}, R_{ajka;n}R_{bjnb;k}, R_{ijk\ell;n}R_{ijk\ell;n},$
$R_{ijji}R_{k\ell\ell k;nn}, R_{ajka}R_{bjkb;nn}, R_{ajka}R_{bjnb;kn}, R_{ijk\ell}R_{ijk\ell;nn}, R_{ijji}R_{k\ell\ell k}R_{abba},$
$R_{ijji}R_{ajka}R_{bjkb}, R_{ijji}R_{abcd}R_{abcd}, R_{ajka}R_{bjnb}R_{cknc}, R_{aija}R_{bk\ell b}R_{ikj\ell},$
$R_{ajka}R_{jn\ell i}R_{kn\ell i}, R_{ijkn}R_{ij\ell p}R_{kn\ell p}, R_{ijkn}R_{i\ell kp}R_{j\ell np}\}$.

The order can also be defined more invariantly in terms of rescaling or homothety. One can use dimensional analysis to establish the following result (see, for example, Gilkey [71]).

Lemma 9.26 Let $Q \in \mathcal{I}_m$. Then $Q \in \mathcal{I}_{m,n}$ if and only if $Q(c^2 g, P) = c^{-n} Q(g, P)$ for all $0 \neq c \in \mathbb{R}$ and all (M, g, P).

9.2.3 UNIVERSAL CURVATURE IDENTITIES.

There is a natural restriction map

$$r : \mathcal{I}_{m,n} \to \mathcal{I}_{m-1,n}$$

given by restricting the summation to range from 1 to $m-1$ rather than from 1 to m. For example, $r(\tau_m) = \tau_{m-1}$. The scalar curvature is a *universal tensor* and for that reason we will not subscript it in this way. The universal scalar invariants given in Lemma 9.25 are linearly independent if $m \geq n$. However, they are not linearly independent if $m = n - 1$ and there is a single additional universal relation amongst these invariants that we may describe as follows. Define the *Pfaffian* $E_{m,n} \in \mathcal{I}_{m,n}$ for n even by setting:

$$E_{m,n} := R_{i_1 i_2 j_2 j_1} \cdot \dots \cdot R_{i_{n-1} i_n j_n j_{n-1}} g(e^{i_1} \wedge \dots \wedge e^{i_n}, e^{j_1} \wedge \dots \wedge e^{j_n}).$$

For example, $E_{m,2} = 2\tau_m$ is essentially just the scalar curvature. The invariants $E_{m,n}$ are, again, *universal*, i.e., $E_{m,n} \in \mathcal{I}_{m,n}$ and $r(E_{m,n}) = E_{m-1,n}$. Clearly, $E_{m,n} = 0$ for $m < n$ since $e^{i_1} \wedge \dots \wedge e^{i_n}$ vanishes. Consequently, $E_{m,m} \in \ker\{r : \mathcal{I}_{m,m} \to \mathcal{I}_{m-1,m}\}$ and $E_{m,m}$ provides a *universal curvature identity*. Expressing the invariants $E_{m,2}$, $E_{m,4}$, and $E_{m,6}$ universally in terms of contractions of indices (see, for example, the discussion in Pekonen [115]) then yields the following result.

Lemma 9.27

1. If $m = 1$, then $0 = R_{ijji}$.
2. If $m = 3$, then $0 = R_{ijji}R_{k\ell\ell k} - 4R_{aija}R_{bijb} + R_{ijk\ell}R_{ijk\ell}$.
3. If $m = 5$, then $0 = R_{ijji}R_{k\ell\ell k}R_{abba} - 12R_{ijji}R_{aija}R_{bijb} + 3R_{abba}R_{ijk\ell}R_{ijk\ell}$
$+24R_{aija}R_{bk\ell b}R_{j\ell ik} + 16R_{aija}R_{bjkb}R_{cikc} - 24R_{aija}R_{jk\ell n}R_{\ell nik}$
$+2R_{ijk\ell}R_{k\ell an}R_{anij} - 8R_{kaij}R_{ink\ell}R_{j\ell an}$.

Lemma 9.27 gives the only such universal relations of this type in dimensions 1, 3, and 5. We postpone the proof of the following more general result until Section 9.2.5 as we will also establish similar results for symmetric 2-tensor-valued invariants. It was first established by Gilkey [71] in his heat equation proof of the Chern–Gauss–Bonnet Theorem.

Theorem 9.28

1. *$r : \mathcal{I}_{m,n} \to \mathcal{I}_{m-1,n}$ is always surjective.*
2. *If n is even and if $m > n$, then $r : \mathcal{I}_{m,n} \to \mathcal{I}_{m-1,n}$ is bijective.*
3. *Let m be even. Then* $\ker\{r : \mathcal{I}_{m,m} \to \mathcal{I}_{m-1,m}\} = E_{m,m} \cdot \mathbb{R}$.

9.2.4 SYMMETRIC 2-TENSOR-VALUED INVARIANTS. Let $\mathcal{I}^2_{m,n}$ be the space of symmetric 2-tensor-valued invariants which are homogeneous of order n in the derivatives of the metric and which are defined in the category of m-dimensional Riemannian manifolds. Let $\{e_1, \dots, e_k\}$ be a local orthonormal frame for the tangent bundle of M. If ξ and η are cotangent vectors, then the symmetric product is denoted by $\xi \circ \eta := \frac{1}{2}\{\xi \otimes \eta + \eta \otimes \xi\}$. For example, $g = e^k \circ e^k$. Theorem 9.23 extends easily to this situation; all invariants are obtained from monomial expressions in the curvature where one symmetrizes two indices and contracts in pairs the remaining indices. After eliminating redundancies, one has:

Lemma 9.29

1. $\mathcal{I}^2_{m,0} = \text{span}\{e^k \circ e^k\}$.
2. $\mathcal{I}^2_{m,2} = \text{span}\{R_{ijji}e^k \circ e^k, R_{ijki}e^j \circ e^k\}$.
3. $\mathcal{I}^2_{m,4} = \text{span}\{R_{ijji;kk}e^\ell \circ e^\ell, R_{kjj\ell;ii}e^k \circ e^\ell, R_{ijji;k\ell}e^k \circ e^\ell,$
$R_{ijji}R_{k\ell\ell k}e^n \circ e^n, R_{ijki}R_{\ell jk\ell}e^n \circ e^n, R_{ijk\ell}R_{ijk\ell}e^n \circ e^n, R_{ijji}R_{k\ell nk}e^\ell \circ e^n,$
$R_{ik\ell i}R_{jknj}e^\ell \circ e^n, R_{ijk\ell}R_{ijkn}e^\ell \circ e^n, R_{\ell ijn}R_{kijk}e^\ell \circ e^n\}$.

Restricting the range of summation and setting $e^j \circ e^k = 0$ if $j = m$ or if $k = m$ yields an analogous restriction map $r : \mathcal{I}^2_{m,n} \to \mathcal{I}^2_{m-1,n}$; the elements given in Lemma 9.29 are universal with respect to restriction. They are linearly independent if $m > n$, but there is a single relation if $m = n$. For n even, define $T^2_{m,n} \in \mathcal{I}^2_{m,n}$ by setting:

$$T^2_{m,n} := R_{i_1 i_2 j_2 j_1} \cdots R_{i_{n-1} i_n j_n j_{n-1}} e^{i_{n+1}} \circ e^{j_{n+1}} g(e^{i_1} \wedge \cdots \wedge e^{i_{n+1}}, e^{j_1} \wedge \cdots \wedge e^{j_{n+1}}). \quad (9.2.a)$$

Clearly, these elements are universal and $T^2_{m,m} = 0$. This then leads to the following identities; we refer as well to Euh, Park and Sekigawa [61, 62] for a proof using a different approach.

Lemma 9.30

1. If $m = 2$, then $0 = R_{ijji}e^k \circ e^k - 2R_{ijki}e^j \circ e^k$.
2. If $m = 4$, then $0 = -\frac{1}{4}\{R_{ijji}R_{k\ell\ell k} - 4R_{ijki}R_{\ell jk\ell} + R_{ijk\ell}R_{ijk\ell}\}e^n \circ e^n$
$+\{R_{k\ell ni}R_{k\ell nj} - 2R_{knik}R_{\ell nj\ell} - 2R_{ik\ell j}R_{nk\ell n} + R_{k\ell\ell k}R_{nijn}\}e^i \circ e^j$.
3. If $m = 6$, then $0 = \frac{1}{2}\{\tau^3 + 3\tau\|R\|^2 - 12\tau\|\rho\|^2 + 16\rho_{ab}\rho_{bc}\rho_{ca} - 24\rho_{ab}\rho_{cd}R_{acbd}$
$-24\rho_{uv}R_{abcu}R_{abcv} + 8R_{dabc}R_{dubv}R_{aucv} - 4R_{dbac}R_{dbuv}R_{acuv}\}e^i \circ e^i$
$+\{-3\tau^2\rho_{ij} - 3\|R\|^2\rho_{ij} + 12\|\rho\|^2\rho_{ij} + 12\tau\rho_{ia}\rho_{ja} + 12\tau\rho_{ab}R_{iabj} - 6\tau R_{iabc}R_{jabc}$
$-24\rho_{ia}\rho_{jb}\rho_{ab} - 24\rho_{ac}\rho_{bc}R_{iabj} + 24\rho_{aj}\rho_{cd}R_{acid} + 24\rho_{ai}\rho_{cd}R_{acjd}$
$+24\rho_{ab}R_{icdj}R_{acbd} + 48\rho_{cd}R_{iabc}R_{jabd} + 6\rho_{jd}R_{abci}R_{abcd} + 6\rho_{id}R_{abcj}R_{abcd}$
$+12R_{iuvj}R_{abcu}R_{abcv} + 12R_{ibac}R_{jbuv}R_{acuv} - 24R_{iabc}R_{jubv}R_{aucv}\}e^i \circ e^j$.

In fact, the identities of Lemma 9.30 are the only universal identities of this form if $m = 2$, $m = 4$, or $m = 6$. We will establish the following result in Section 9.2.5.

Theorem 9.31

1. *$r : \mathcal{I}^2_{m,n} \to \mathcal{I}^2_{m-1,n}$ is always surjective.*
2. *If n is even and if $m > n + 1$, then $r : \mathcal{I}^2_{m,n} \to \mathcal{I}^2_{m-1,n}$ is bijective.*
3. *If m is even, then* $\ker\{r : \mathcal{I}^2_{m+1,m} \to \mathcal{I}^2_{m,m}\} = T^2_{m+1,m} \cdot \mathbb{R}$.

Example. We have $T^2_{m,2} = R_{i_1 i_2 j_2 j_1} e^{i_3} \circ e^{j_3} g(e^{i_1} \wedge e^{i_2} \wedge e^{i_3}, e^{j_1} \wedge e^{j_2} \wedge e^{j_3})$. Set $m = 2$. Then $0 = T^2_{2,2} = 2R_{ijji}e^k \circ e^k - 4R_{kijk}e^i \circ e^j$ so $\rho = \frac{1}{2}\tau_2 g$ on any Riemann surface which is a well-known identity.

9.2.5 THE PROOF OF THEOREM 9.28 AND THEOREM 9.31. The first Assertion of Theorem 9.28 and Theorem 9.31 follow from the following observation:

Lemma 9.32 r defines a surjective map from $\mathcal{I}_{m,n}$ to $\mathcal{I}_{m-1,n}$ and from $\mathcal{I}^2_{m,n}$ to $\mathcal{I}^2_{m-1,n}$.

Proof. By Lemma 9.24, all local invariants are given in terms of contractions of indices of various monomials of weight n in the covariant derivatives of the curvature tensor. Instead of letting the indices range from 1 to m in the contractions of indices which define Q, we let the indices range from 1 to $m-1$ in defining $r(Q)$ since the metric is flat in the last direction. We may choose a spanning set for $\mathcal{I}_{m-1,n}$ or $\mathcal{I}^2_{m-1,n}$ similar to those given in Lemma 9.25 and Lemma 9.29 which involves contracting indices in covariant derivatives of the curvature tensor. The desired lift to $\mathcal{I}_{m,n}$ or to $\mathcal{I}^2_{m,n}$ is then obtained by letting the indices range from 1 to m instead of from 1 to $m-1$. This lift is, of course, not unique and this indeterminacy is measured by $\ker\{r\}$ which gives the universal relations satisfied in dimension $m-1$ which are not satisfied in dimension m. □

To prove the remaining assertions of Theorem 9.28 and Theorem 9.31, we introduce a different formalism. Instead of working with the covariant derivatives of the curvature tensor and dealing with the Bianchi identities, it is convenient to work non-invariantly with the derivatives of the metric. Fix the underlying dimension m. Let $\alpha = (a_1, \dots, a_m)$ be a multi-index of order $|\alpha| := a_1 + \dots + a_m$. Introduce formal variables $g_{ij/\alpha}$ for the derivatives of the metric. Let P be a point of a Riemannian manifold (M, g). We can always normalize the coordinate systems so that $g_{ij}(\vec{x}, P) = \delta_{ij}$ and so that $g_{ij/k}(\vec{x}, P) = 0$; such a coordinate system is said to be an *admissible coordinate system*. By restricting to admissible coordinate systems, we may assume that $|\alpha| \geq 2$. Let $\mathcal{Q}_m$ be the free commutative unital real algebra generated by the variables $g_{ij/\alpha}$ for $|\alpha| \geq 2$. We define $Q(\vec{x}, g, P) \in \mathbb{R}$ by substitution in the obvious fashion and say that Q is *invariant* if $Q(\vec{x}, g, P)$ is independent of the coordinate system $\vec{x}$ for every possible admissible coordinate system and denote this common value by $Q(g, P)$. Let $\mathcal{I}_{m,n}$ be the vector space of all such invariant local formulae which are homogeneous of order n. More generally, instead of considering scalar functions, we could consider symmetric 2-tensor-valued invariants

$$Q = Q_{ij} dx^i \circ dx^j \quad \text{for} \quad Q_{ij} \in \mathcal{Q}_m;$$

we will say that Q is invariant if $Q(\vec{x}, g, P)$ is independent of $\vec{x}$ for all (g, P).

We can now describe the restriction map. Let (N, g_N) be a Riemannian manifold of dimension $m-1$. Let $M = N \times S^1$ and let $g_M = g_N + d\theta^2$ where θ is the usual periodic parameter on the circle. Let θ_0 be the basepoint of the circle; since $(S^1, d\theta^2)$ is a homogeneous space, the choice of the basepoint plays no role. Define the inclusion i from N to M by setting $i(y) := (y, \theta_0) \in M$. If $Q \in \mathcal{I}_{m,n}$ or if $Q \in \mathcal{I}^2_{m,n}$, then restrict to $N \times \{\theta_0\}$ and define:

$r(Q)(g_N, y) := i^* Q(g_M, i(y))$. Dually, this defines the restriction maps $r : \mathcal{I}_{m,n} \to \mathcal{I}_{m-1,n}$ and $r : \mathcal{I}^2_{m,n} \to \mathcal{I}^2_{m-1,n}$. The following result can be proved using Taylor series.

Lemma 9.33 If $0 \neq Q \in \mathcal{Q}_m$, there exists a point P of some Riemannian manifold (M, g) and an admissible coordinate system $\vec{x}$ so that $Q(\vec{x}, g, P) \neq 0$.

We note that Lemma 9.33 is **not** true if we work with the Riemann curvature tensor. There are "hidden" and non-obvious relations that do not follow from the usual $\mathbb{Z}_2$-symmetries and the generalized Bianchi identities that are dimension specific – that is the whole point, of course, of the relations given in Lemma 9.27 and Lemma 9.30. And it is Lemma 9.33 that will be crucial in our discussion. Let $A = g_{i_1 j_1/\alpha_1} \cdot \cdots \cdot g_{i_\ell j_\ell/\alpha_\ell}$ be a monomial of $\mathcal{Q}_m$. We let $\deg_k(A) := \delta_{i_1,k} + \delta_{j_1,k} + \alpha_1(k) + \cdots + \delta_{i_\ell,k} + \delta_{j_\ell,k} + \alpha_\ell(k)$ be the number of times that the index k appears in A. We extend this notion to symmetric 2-tensor-valued invariants by defining $\deg_k(A\, dx^{i_{\ell+1}} \circ dx^{j_{\ell+1}}) := \deg_k(A) + \delta_{i_{\ell+1},k} + \delta_{j_{\ell+1},k}$. It is then obvious that the restriction map r defined previously takes the form:

$$r(A) = \left\{ \begin{array}{ll} A & \text{if } \deg_m(A) = 0 \\ 0 & \text{otherwise} \end{array} \right\} .$$

Assertion 2 of Theorem 9.28 and Theorem 9.31 are consequences of the following result.

Lemma 9.34 Let A be a monomial of Q.

1. If $Q \in \mathcal{I}_{m,n} \cap \ker\{r\}$ or if $Q \in \mathcal{I}^2_{m,n} \cap \ker\{r\}$, then

$$\deg_k(A) \geq 2 \quad \text{for} \quad 1 \leq k \leq m .$$

2. If $m > n$, then $\ker\{r : \mathcal{I}_{m,n} \to \mathcal{I}_{m-1,n}\} = \{0\}$.
3. If $m = n$ and $Q \in \mathcal{I}_{m,n} \cap \ker\{r\}$, then

$$\deg_k(A) = 2 \quad \text{for} \quad 1 \leq k \leq m \quad \text{and} \quad |\alpha_a| = 2 \quad \text{for} \quad 1 \leq a \leq \ell .$$

4. If $m > n + 1$, then $\ker\{r : \mathcal{I}^2_{m,n} \to \mathcal{I}^2_{m-1,n}\} = \{0\}$.
5. If $m = n + 1$ and $Q \in \mathcal{I}^2_{m,n} \cap \ker\{r\}$, then

$$\deg_k(A) = 2 \quad \text{for} \quad 1 \leq k \leq m \quad \text{and} \quad |\alpha_a| = 2 \quad \text{for} \quad 1 \leq a \leq \ell .$$

Proof. Let $r(Q) = 0$. By Lemma 9.33, we may identify the local formula defined by Q with the polynomial $Q \in \mathcal{Q}_m$. It then follows that $\deg_m(A) > 0$ for every monomial A of Q. By considering the change of coordinates $(x^1, \ldots, x^{m-1}, x^m) \to (x^1, \ldots, x^{m-1}, -x^m)$ and by using the fact that Q is invariant, we see that $\deg_m(A)$ is even and, consequently, $\deg_m(A) \geq 2$. Since

Q is invariant under coordinate permutations, Assertion 1 follows. Let $0 \neq Q \in \mathcal{I}_{m,n}$ satisfy $r(Q) = 0$. Let $A = g_{i_1 j_1/\alpha_1} \cdot \dots \cdot g_{i_\ell j_\ell/\alpha_\ell}$ be a monomial of Q. We have $|\alpha_a| \geq 2$. Furthermore, since $r(Q) = 0$, $\deg_k(A) \geq 2$ for every k. We estimate:

$$\begin{aligned} 2\ell &\leq |\alpha_1| + \dots + |\alpha_\ell| = n, \\ 2m &\leq \sum_{k=1}^{m} \deg_k(A) = \sum_{a=1}^{\ell} \sum_{k=1}^{m} \{\delta_{i_a,k} + \delta_{j_a,k} + \alpha_a(k)\} \\ &= \sum_{a=1}^{\ell} \{1 + 1 + |\alpha_a|\} = 2\ell + n \leq n + n = 2n\,. \end{aligned}$$

This shows $m \leq n$ and proves Assertion 2. Furthermore, if $m = n$, all these inequalities must have been equalities; Assertion 3 now follows. Similarly, let $0 \neq Q \in \mathcal{I}^2_{m,n} \cap \ker\{r\}$ and let A be a monomial of Q. Express $A = g_{i_1 j_1/\alpha_1} \cdots g_{i_\ell j_\ell/\alpha_\ell} dx^{i_{\ell+1}} \circ dx^{j_{\ell+1}}$. We estimate similarly:

$$\begin{aligned} &2\ell \leq |\alpha_1| + \dots + |\alpha_\ell| = n, \\ &2m \leq \sum_{k=1}^{m} \deg_k(A) = 2 + \sum_{a=1}^{\ell} \sum_{k=1}^{m} \{\delta_{i_a,k} + \delta_{j_a,k} + \alpha_a(k)\} \\ &\quad = 2 + \sum_{a=1}^{\ell} \{1 + 1 + |\alpha_a|\} = 2 + 2\ell + n \leq 2 + 2n\,. \end{aligned}$$

Again, this is not possible if $m > n + 1$ which establishes Assertion 4. If $m = n + 1$, all the equalities must have been equalities and Assertion 5 follows. □

We now establish Assertion 3 of Theorem 9.28. Let $m = 2\bar{m}$ be even. We introduce formal variables $g_{ij/k\ell} \in S^2 \otimes S^2$ for $1 \leq i, j, k, \ell \leq m$. If $Q \in \ker\{r : \mathcal{I}_{m,m} \to \mathcal{I}_{m-1,m}\}$, then we have shown in Lemma 9.34 that Q can be regarded as a polynomial of degree $\bar{m}$ in $\mathbb{R}[g_{ij/k\ell}]$. Let S^2 denote the space of symmetric 2-tensors. Since $g_{ij/k\ell} \in S^2 \otimes S^2$, we can regard Q as a linear orthogonal invariant on $\otimes^{\bar{m}}\{S^2 \otimes S^2\} \subset \otimes^{2m}(V)$. Such an orthogonal invariant extends naturally to the full tensor algebra to be zero on the orthogonal complement of $\otimes^{\bar{m}}\{S^2 \otimes S^2\}$. Since the restriction of Q to the lower-dimensional setting vanishes, we can apply Assertion 1 of Theorem 9.23 to express Q as a linear combination of invariants of the form

$$\begin{aligned} A_\sigma = \; & g_{i_1 i_2/i_3 i_4} \cdots g_{i_{2m-3} i_{2m-2}/i_{2m-1} i_{2m}} \\ & \times g(dx^{i_{\sigma_1}} \wedge dx^{i_{\sigma_2}} \wedge \dots \wedge dx^{i_{\sigma_m}}, dx^{i_{\sigma_{m+1}}} \wedge \dots \wedge dx^{i_{\sigma_{2m}}}), \end{aligned}$$

where σ is a permutation of $\{1, \dots, 2m\}$. If $1 = \sigma_a$ for some index a with $1 \leq a \leq m$, then necessarily $2 = \sigma_b$ for some index b with $m + 1 \leq b \leq 2m$ since $g_{i_1 i_2/i_3 i_4}$ is symmetric in the indices $\{i_1, i_2\}$ whereas the wedge product is anti-symmetric. By interchanging the indices $\{i_1, i_2\}$ if necessary, we may therefore assume $i_1 = \sigma_{a_1}$ and $i_2 = \sigma_{b_1}$ for $1 \leq a_1 \leq m$ and $m + 1 \leq b_1 \leq 2m$.

This implies that we can write

$$\begin{aligned}A_\sigma &= g_{i_1 j_1/i_2 j_2}\cdot\dots\cdot g_{i_{m-1}j_{m-1}/i_m j_m}\\&\quad\times g(dx^{i_{\rho_1}}\wedge\dots\wedge dx^{i_{\rho_m}}, dx^{j_{\varrho_1}}\wedge\dots\wedge dx^{j_{\varrho_m}})\end{aligned}$$

where ρ and ϱ are permutations of m indices. Reordering the factors then yields

$$\begin{aligned}A_\sigma &= \pm g_{i_1 j_1/i_2 j_2}\cdot\dots\cdot g_{i_{m-1}j_{m-1}/i_m j_m}\\&\quad\times g(dx^{i_1}\wedge\dots\wedge dx^{i_m}, dx^{j_1}\wedge\dots\wedge dx^{j_m})\,.\end{aligned}$$

This shows $\dim(\ker\{r:\mathcal{I}_{m,m}\to\mathcal{I}_{m-1,m}\})\le 1$. Since $r(E_{m,m})=0$ and $E_{m,m}$ is non-trivial, Assertion 3 of Theorem 9.28 follows. □

The proof of Assertion 3 of Theorem 9.31 is essentially the same. The crucial feature is, of course, that we have eliminated the higher-order jets of the metric and only have to deal with second derivatives. Let $Q\in\mathcal{I}^2_{m+1,m}$. We can express $Q=Q_{uv}dx^u\circ dx^v$ where $Q_{uv}\in\mathbb{R}[g_{ij/k\ell}]$ is homogeneous of degree $\bar{m}$. Since $r(Q)=0$, we may express Q as a linear combination of invariants of the form:

$$\begin{aligned}A_\sigma &= g_{i_1 i_2/i_3 i_4}\cdots g_{i_{2m-3}i_{2m-2}/i_{2m-1}i_{2m}}dx^{i_{2m+1}}\circ dx^{i_{2m+2}}\\&\quad\times g(dx^{i_{\sigma_1}}\wedge dx^{i_{\sigma_2}}\wedge\dots\wedge dx^{i_{\sigma_{m+1}}}, dx^{i_{\sigma_{m+2}}}\wedge\dots\wedge dx^{i_{\sigma_{2m+2}}})\,.\end{aligned}$$

The same symmetry argument used to establish Theorem 9.28 then shows in fact we are dealing with

$$\begin{aligned}A_\sigma &= \pm g_{i_1 j_1/j_2 i_2}\cdots g_{i_{m-1}j_{m-1}/j_m i_m}dx^{i_{m+1}}\circ dx^{j_{m+1}}\\&\quad\times g(dx^{i_1}\wedge dx^{i_2}\wedge\dots\wedge dx^{i_{m+1}}, dx^{j_1}\wedge dx^{j_2}\wedge\dots\wedge dx^{j_{m+1}})\,.\end{aligned}$$

Again, this shows $\dim(\ker\{r:\mathcal{I}^2_{m+1,m}\to\mathcal{I}^2_{m,m}\})\le 1$. The desired result then follows as $T^2_{m+1,m}\in\ker\{r:\mathcal{I}^2_{m+1,m}\to\mathcal{I}^2_{m,m}\}$ is non-trivial. □

9.3 THE CHERN–GAUSS–BONNET FORMULA

The Pfaffian $E_{m,m}$ is important not only in invariance theory, but also because it is the integral of the Chern–Gauss–Bonnet Theorem. In Section 9.3.1, we determine the Euler–Lagrange equations associated with the Chern–Gauss–Bonnet Theorem. In Section 9.3.2, we use this result to extend the Chern–Gauss–Bonnet Theorem from the Riemannian to the pseudo-Riemannian context. Section 9.3.3 presents some examples checking the relevant signs and Section 9.3.4 examines the setting of manifolds with boundary.

Let $\chi(M^2)$ be the *Euler characteristic* of a compact smooth 2-dimensional Riemannian manifold without boundary and let $\tau := R_{ijji}$ be the *scalar curvature*. The classical 2-dimensional *Gauss–Bonnet formula* relates the total scalar curvature of M^2 with $\chi(M^2)$:

$$\chi(M^2)=\frac{1}{4\pi}\int_{M^2}\tau\,dv_g\,. \tag{9.3.a}$$

This formula has been generalized to the higher-dimensional setting by Chern [44] (see related work by Allendoerfer and Weil [5]). For example, if M^4 is a compact 4-dimensional Riemannian manifold without boundary, one has:

$$\chi(M^4) = \frac{1}{32\pi^2}\int_{M^4}\{\tau^2 - 4\|\rho\|^2 + \|R\|^2\}dv_g\,.$$

S. Chern (1911–2004) C. F. Gauss (1777–1855) P. Bonnet (1819–1892)

We refer to Chern [44] for the proof of the following result which generalizes the formulas given above. There is also a heat equation proof due to Gilkey [71] and Patodi [113]. Note that the Euler characteristic $\chi(M)$ of any compact manifold without boundary of odd dimension vanishes so only the even-dimensional case is of interest.

Theorem 9.35 *Let (M, g) be a compact Riemannian manifold without boundary which has dimension $m = 2\ell$. Let $\chi(M)$ be the Euler–Poincaré characteristic. Then*

$$\chi(M) = \int_M \frac{1}{(8\pi)^\ell \ell!}E_{m,m}(g)dv_g\,.$$

9.3.1 EULER–LAGRANGE EQUATIONS. Let g_ε be a smooth 1-parameter family of metrics with $g(0) = g$. Set $h := \partial_\varepsilon g_\varepsilon|_{\varepsilon=0}$. Since the Euler form $E_{m,n}$ only involves the first and second derivatives of the metric, the variation only involves the first and second derivatives of h. Let $h_{ij;k}$ and $h_{ij;k\ell}$ give the components of the first and second covariant derivatives of h with respect to the Levi–Civita connection of g. We express

$$\partial_\varepsilon\left\{E_{m,n}(g_\varepsilon)dv_{g_\varepsilon}\right\}\big|_{\varepsilon=0} = Q^{ij}_{m,n}h_{ij} + Q^{ijk}_{m,n}h_{ij;k} + Q^{ijk\ell}_{m,n}h_{ij;k\ell}\,.$$

Let $Q^{ijk}_{m,n,\ell}$ and $Q^{ijk\ell}_{m,n,uv}$ be the components of the first and second covariant derivatives of these tensors, respectively. Define:

$$S^2_{m,n} := \{Q^{ij}_{m,n} - Q^{ijk}_{m,n,k} + Q^{ijk\ell}_{m,n,\ell}\}e^i \circ e^j\,.$$

This tensor is characterized by the property that if (M, g) is any compact Riemannian manifold without boundary of dimension m, then we may integrate by parts to see that:

$$\partial_\varepsilon\left\{\int_M E_{m,n}(g_\varepsilon)dv_{g_\varepsilon}\right\}\bigg|_{\varepsilon=0} = \int_M S^2_{m,n,ij}h_{ij}dv_g\,.$$

These are the Euler–Lagrange equations of Chern–Gauss–Bonnet gravity; we refer to the discussion in Lovelock [103] for further details in this regard. In the following result, we establish a conjecture of Berger [12] that $S^2_{m,n}$ involves only the first and second derivatives of the metric and not, as is a priori possible, the 3rd and 4th derivatives of the metric, i.e., the 1st and 2nd covariant derivatives of the curvature tensor. This result is, of course, not new. It was first established by Kuz'mina [95] and subsequently established using different methods by Labbi [96, 97, 98]. It illustrates the close connection between index theory and invariance theory.

Theorem 9.36 *If g_ε is a smooth 1-parameter family of Riemannian metrics on a compact manifold M of dimension m, then*

$$\partial_\varepsilon \left\{ \int_M E_{m,2\ell}(g_\varepsilon) dv_{g_\varepsilon} \right\}\Big|_{\varepsilon=0} = \int_M \langle T^2_{m,2\ell}, h\rangle dv_g .$$

Proof. Since both sides vanish identically if $m \le 2\ell$, we assume $m > 2\ell$. It is immediate from the definition that $S^2_{m,2\ell} \in \mathcal{I}^2_{m,2\ell}$ and also that $r(S^2_{m,2\ell}) = S^2_{m-1,2\ell}$. Theorem 9.35 shows $S^2_{2\ell,2\ell} = 0$ so $S^2_{2\ell+1,2\ell}$ belongs to the kernel of r mapping $\mathcal{I}^2_{2\ell+1,2\ell}$ to $\mathcal{I}^2_{2\ell,2\ell}$. Thus, by Theorem 9.31, there exists a universal constant d_ℓ so $S^2_{2\ell+1,2\ell} = d_\ell T^2_{m,2\ell}$. Let $m \ge 2\ell + 2$. We apply Theorem 9.31 once again to see that r is a bijective map from $\mathcal{I}^2_{m,2\ell}$ to $\mathcal{I}^2_{m-1,2\ell}$. It is immediate that

$$r(S^2_{m,2\ell}) = S^2_{m,2\ell} \quad \text{and} \quad r(T^2_{m,2\ell}) = T^2_{m,2\ell} .$$

Thus $S^2_{m,2\ell} = d_\ell T^2_{m,2\ell}$ for any $m > 2\ell$.

Since both sides vanish identically if $m \le 2\ell$, the restriction $m > 2\ell$ is unnecessary. Thus, to complete Theorem 9.36, it suffices to show $d_\ell = 1$. We may take $m = 2\ell + 1$. Choose a Riemannian manifold (N, g_N) of dimension 2ℓ with $\chi(N) \ne 0$. Theorem 9.35 then yields:

$$\int_M E_{2\ell,2\ell}(g_N) dv_{g_N} \ne 0 .$$

Let $M := N \times S^1$ and let $g_\varepsilon := g_N + e^{2\varepsilon} d\theta^2$. Then $E_{2\ell+1,2\ell}(g_\varepsilon)$ is independent of ε while $dv_{g_\varepsilon} = e^\varepsilon v_g$. Consequently,

$$\partial_\varepsilon \left\{ \int_M E_{m,2\ell}(g_\varepsilon) dv_{g_\varepsilon} \right\}\Big|_{\varepsilon=0} = 2\pi \cdot \partial_\varepsilon \{e^\varepsilon\}|_{\varepsilon=0} \cdot \int_N E_{2\ell,2\ell}(g_N) dv_{g_N} .$$

We set $n = 2\ell$ in Equation (9.2.a). Since the metric is flat in the S^1 direction, we must take $e^{i_{n+1}} = e^{j_{n+1}} = d\theta$. This implies that $\langle T^2_{2\ell+1,2\ell}(g), h\rangle = E_{2\ell,2\ell}(g_N)$ so

$$\int_M \langle T^2_{2\ell+1,2\ell}(g), h\rangle dv_g = 2\pi \cdot \int_N E_{2\ell,2\ell}(g_N) dv_{g_N} .$$

The desired identity $d_\ell = 1$ now follows from the preceding identities. □

In preparation for our discussion of the Chern–Gauss–Bonnet Theorem in the pseudo-Riemannian context, it is convenient to pass to the complex setting. We suppose given a complex metric $g \in C^\infty(S^2(T^*M) \otimes \mathbb{C})$ and assume $\det(g) \neq 0$ as a non-degeneracy condition. The Levi–Civita connection and curvature tensor may then be defined. To maintain analyticity, we set $\mathrm{d}v_g := \sqrt{\det(g_{ij})}\, dx^1 \cdot \dots \cdot dx^m$; we do not take the absolute value. There is a subtlety here since, of course, there are two branches of the square root function but we ignore this topological difficulty for the moment.

Lemma 9.37 Let M be a compact manifold of dimension m. Let g_ε be a smooth 1-parameter family of complex metrics on M so that $\sqrt{\det(g_\varepsilon)}$ can be defined consistently on M for ε in the parameter range. Let $h_\varepsilon = \partial_\varepsilon g_\varepsilon$. Then

$$\partial_\varepsilon \left\{ \int_M E_{m,n}(g_\varepsilon)\, \mathrm{d}v_{g_\varepsilon} \right\}\Bigg|_{\varepsilon=0} = \int_M \frac{1}{(8\pi)^\ell \ell!} \langle T^2_{m,n}(g_0), h_0 \rangle dv_{g_0} \,.$$

If the family contains a Riemannian metric and if $m = 2\ell$, then

$$\int_M E_{m,n}(g_\varepsilon)\, \mathrm{d}v_{g_\varepsilon} = \chi(M) \quad \text{for any} \quad \varepsilon.$$

Proof. We regard $E_{m,n}$, $T^2_{m,n}$, the curvature tensor R, the covariant derivative of the curvature ∇R, and so forth as polynomials in the derivatives of the metric tensor with coefficients which are analytic in the g_{ij} variables. The first identity is then an identity between two analytic expressions in the variables $\{g_{ij}, g_{ij/k}, g_{ij/k\ell}, \dots\}$ where $\det(g_{ij}) \neq 0$. Since the zeros of $\det(g_{ij})$ have complex codimension 1 in the linear space of symmetric 2-tensors, the condition $\det(g_{ij}) \neq 0$ does not disconnect the parameter space. By Theorem 9.36, the first identity holds where $\det(g_{ij}) \neq 0$, g_{ij} is real, and the signature is positive definite. Thus, by analytic continuation, it holds in general. If we set $n = m$, then $T^2_{m,m} = 0$ so $\int_M E_{m,m}(g_\varepsilon)\, \mathrm{d}v_{g_\varepsilon}$ is independent of ε. The second identity now follows from Theorem 9.35 if the family contains a Riemannian metric. □

9.3.2 THE GENERALIZED CHERN–GAUSS–BONNET THEOREM.

Avez [9] and Chern [46] independently extended Theorem 9.35 to the indefinite setting (there is a slight mistake in the original paper of Chern [46] as the the sign change $(-1)^{p/2}$ is not present).

Let (M, g) be a compact pseudo-Riemannian manifold without boundary of signature (p, q). Recall that the volume element is given by

$$dv_g = \sqrt{|\det(g_{ij})|}\, dx^1 \cdot \dots \cdot dx^m \,.$$

Theorem 9.38 *Let (M, g) be a compact pseudo-Riemannian manifold of signature (p, q) without boundary of even dimension m. If p is odd, then $\chi(M)$ vanishes. If p is even, then*

$$\chi(M) = (-1)^{p/2} \int_M E_{m,m} dv_g .$$

Proof. We use Lemma 9.37. Let (M, g_1) be a compact pseudo-Riemannian manifold without boundary of even dimension m and signature (p, q). Apply Lemma 9.3 to decompose the tangent bundle $TM = V_- \oplus V_+$. Let $g_\pm := \pm g_1|_{V_\pm}$ so $g_1 = -g_- \oplus g_+$. Let $g_0 := g_- \oplus g_+$ be the corresponding Riemannian metric on M. We follow a circular arc from 1 to -1 in the complex plane given by $e^{\varepsilon\pi\sqrt{-1}}$ for $\varepsilon \in [0, \pi]$ to define a variation connecting g_0 to g_1:

$$g_\varepsilon := e^{\varepsilon\pi\sqrt{-1}} g_- \oplus g_+ . \tag{9.3.b}$$

We note $\det(g_\varepsilon) = \det(g_0) e^{p\varepsilon\pi\sqrt{-1}}$ so the family is admissible and we have:

$$\sqrt{\det(g_\varepsilon)} = e^{p\varepsilon\pi\sqrt{-1}/2} \sqrt{det(g_0)} .$$

If p is odd, then $\sqrt{\det(g_1)}$ will be purely imaginary. Consequently, $\chi(M)$ will be purely imaginary. Since $\chi(M)$ is real, we conclude $\chi(M) = 0$ in this case. On the other hand, if p is even, then $\sqrt{\det(g_1)} = (-1)^{p/2}\sqrt{\det(g_0)}$. Therefore, Theorem 9.38 follows from Lemma 9.37. □

9.3.3 EXAMPLES. Suppose (M, g_0) is a Riemann surface. Let $g_1 = -g_0$ have signature $(2, 0)$. Then the Levi–Civita connection of g_1 and the Levi–Civita connection of g_0 agree so

$$R_{ijk}{}^{\ell}(g_1) = R_{ijk}{}^{\ell}(g_0) \quad \text{and} \quad \tau(g_1) = g_1^{jk} R_{ijk}{}^{i}(g_1) = -g_0^{jk} R_{ijk}{}^{i}(g_0) = -\tau(g_0) .$$

As $dv_{g_0} = dv_{g_1}$, one must change the sign in the Gauss–Bonnet Theorem:

$$\chi(M) = -\frac{1}{4\pi} \int_M \tau(g_1) dv_{g_1} .$$

If $(M, g) = (M_1, h_1) \times (M_2, h_2)$ is the product of two Riemann surfaces, then the Gauss–Bonnet Theorem decouples and we have $\chi(M) = \chi(M_1)\chi(M_2)$ and $E_4(g) = E_2(h_1)E_2(h_2)$. Thus, we will not need to change the sign in signature $(4, 0)$ or $(0, 4)$ but we will need to change the sign in signature $(2, 2)$. The fact that the Euler characteristic vanishes if p and q are both odd is not, of course, new but follows from standard characteristic class theory.

9.3.4 RIEMANNIAN MANIFOLDS WITH BOUNDARY. If M is a 2-dimensional manifold with smooth boundary, then Equation (9.3.a) must be adjusted to include a boundary contribution. Let κ_g be the geodesic curvature. We then have

$$\chi(M) = \frac{1}{4\pi} \int_M \tau dv_g + \frac{1}{2\pi} \int_{\partial M} \kappa_g dv_{g|\partial M} .$$

Chern's original paper [44] also gives a formula for the Euler characteristic in the context of Riemannian manifolds of dimension m with smooth boundary. Let $\{e_1, \ldots, e_m\}$ be a local frame for $TM|_{\partial M}$ where $e_1 \perp T(\partial M)$ and where $e_1 \perp e_a$ for $a \geq 2$. The components of the second fundamental form are then given by

$$L_{ab} := g(e_1, e_1)^{-1/2} g(\nabla_{e_a} e_b, e_1) \quad \text{for} \quad 2 \leq a, b \leq m\,. \tag{9.3.c}$$

Definition 9.39 The *transgression* of the Pfaffian is defined by summing over indices a_i and b_j which range from 2 to m:

$$TE_{m,m}(g) := \sum_{\mu} \left\{ \frac{R_{a_1 a_2 b_2 b_1} \cdot \dots \cdot R_{a_{2\mu-1} a_{2\mu} b_{2\mu} b_{2\mu-1}} L_{a_{2\mu+1} b_{2\mu+1}} \cdot \dots \cdot L_{a_{m-1} b_{m-1}}}{(8\pi)^{\mu} \mu! \operatorname{Vol}(S^{m-1-2\mu})(m-1-2\mu)!} \right.$$
$$\left. \times\, g(e^{a_1} \wedge \dots \wedge e^{a_{m-1}}, e^{b_1} \wedge \dots \wedge e^{b_{m-1}}) \right\}.$$

If m is odd, then $\chi(M) = \frac{1}{2}\chi(\partial M)$ so we may apply Theorem 9.35 to compute $\chi(\partial M)$ and thereby express $\chi(M)$ in terms of curvature. We therefore assume m is even.

Theorem 9.40 *Let (M, g) be a compact smooth manifold Riemannian manifold of even dimension m with smooth boundary ∂M.*

$$\chi(M) = \int_M E_{m,m}(g) dv_g + \int_{\partial M} TE_{m,m}(g) dv_{g|_{\partial M}}\,.$$

Alty [6] generalized this result to the case of pseudo-Riemannian manifolds with boundary under the assumption that the normal vector was either spacelike, timelike, or null on each boundary component by combining the analysis of Avez [9] with the original discussion of Chern [44]. We will not deal with the null case and in the interests of simplicity will simply assume the normal vector to be either timelike or spacelike or, equivalently, that the restriction of the metric to the boundary is non-degenerate. We use Definition 9.39 to define $TE_{m,m}$ in this setting.

Theorem 9.41 *Let (M, g) be a compact smooth pseudo-Riemannian manifold of even dimension m and signature (p, q) which has smooth boundary ∂M. Assume $g|_{\partial M}$ is non-degenerate. If p is odd, then $\chi(M) = 0$. If p is even, then*

$$\chi(M) = (-1)^{p/2} \left\{ \int_M E_{m,m}(g) dv_g + \int_{\partial M} TE_{m,m}(g) dv_{g|_{\partial M}} \right\}.$$

Proof. We use analytic continuation to extend Theorem 9.38 to the pseudo-Riemannian setting rather than, as was done by Alty [6], redo the analysis of Chern in the pseudo-Riemannian context by examining the index of vector fields with isolated singularities. Let (M, g_1) be a pseudo-Riemannian manifold. We suppose $g_1|_{\partial M}$ is non-degenerate. Choose a non-zero vector field X which is normal to the boundary and inward pointing. We can identify a neighborhood of the boundary ∂M in M with $[0, \epsilon) \times \partial M$ and choose local coordinates $(x^1, \ldots, x^m)$ so that $X = \partial_{x^1}$ and so that $\partial M = \{x : x^1 = 0\}$. We then have

$$\det(g_1|_{\partial M}) g_1(X, X)|_{\partial M} = \det(g_1)|_{\partial M} . \tag{9.3.d}$$

We use Lemma 9.3 to choose smooth complementary subbundles $V_\pm$ of TM so that $TM = V_- \oplus V_+$, so that V_+ is perpendicular to V_-, so that the restriction of g_1 to V_+ is positive definite, and so that the restriction of g_1 to V_- is negative definite. We may further normalize the splitting to assume that if the normal vector X is spacelike, then $X \in C^\infty(V^+|_{\partial M})$ while if the normal vector X is timelike, then $X \in C^\infty(V_-|_{\partial M})$. Thus, the splitting $TM = V_- \oplus V_+$ induces a corresponding splitting $T(\partial M) = W_- \oplus W_+$ where $W_\pm = T(\partial M) \cap V_\pm$.

We now consider the smooth 1-parameter of complex variations g_ϵ given above in Equation (9.3.b). The unit normal is then given by $g_\epsilon(X, X)^{-1/2} \cdot X$. In the expressions for $TE_{m,m}$, there are an odd number of terms which contain the second fundamental form L and, consequently, $g_\epsilon(X, X)^{-1/2}$ appears. By Equation (9.3.d):

$$\left\{g_\epsilon(X, X) \det(g_\epsilon|_{\partial M})\right\}^{1/2} = \left\{\det(g_\epsilon)^{1/2}\right\}\Big|_{\partial M} .$$

Thus once again, we must take the square root of $(-1)^p$ in the analytic continuation. Apart from this, the remainder of the argument is the same as that used to prove Theorem 9.38 and, consequently, is omitted. □

9.4 PSEUDO-KÄHLER MANIFOLDS

The material of this section arises from work of Gilkey, Park and Sekigawa [78] and Park [112]. Let $c_{\bar{m}}$ be the $\bar{m}^{\text{th}}$ Chern form (see Chern [45]). If (M, g, J) is a Kähler manifold, then $E_{m,m}(g) dv_g = c_{\bar{m}}(g)$ so this particular characteristic class reproduces the Euler form. The theory of characteristic classes is, of course, much more general and plays an important role in the Hirzebruch–Riemann–Roch Theorem [81]. Thus, the results of this section will in a certain sense generalize Theorem 9.38 to the pseudo-Kähler setting.

Let V be a real vector bundle of dimension 2ℓ which is equipped with an almost complex structure J, i.e., an endomorphism of V with $J^2 = -\text{Id}$. We use J to give V the structure of a complex vector bundle V_c by defining $\sqrt{-1}v := Jv$. Let ∇ be a connection on V which commutes with J. Since J then commutes with the curvature R of ∇, we may regard R as a complex 2-form-valued endomorphism R_c of V_c. Let $\mathfrak{C}_{k,\ell}$ be the collection of polynomial maps

$\Theta(\cdot)$ from the space of $\ell \times \ell$ complex matrices $M_\ell(\mathbb{C})$ to $\mathbb{C}$ which are homogeneous of degree k and which satisfy $\Theta(gAg^{-1}) = \Theta(A)$ for all $A \in M_\ell(\mathbb{C})$ and all g in the general linear group $GL_\ell(\mathbb{C})$. If $\Theta \in \mathfrak{C}_{k,\ell}$, we may define $\Theta(R_c) \in \Lambda^{2k}(M) \otimes_{\mathbb{R}} \mathbb{C}$ invariantly (i.e., to be independent of the particular local frame which was chosen for V_c). One has that $\Theta(R_c)$ is a closed $2k$-form and the de Rham cohomology class of $\Theta(R_c)$ is independent of the particular connection chosen; these are the celebrated *characteristic classes* of Chern [45]. The first Chern class is defined by taking $\Theta(A) := \operatorname{Tr}(A)$ and was discussed in Section 5.4.2 of Book II; we refer to Gilkey [72] for further information concerning the Chern classes noting that there are many excellent references on this subject. Other structure groups, of course, give rise appropriate characteristic classes; the Pontrjagin classes, for example, relate to the orthogonal group while the Euler form which was discussed previously can properly be regarded as a characteristic class of the special orthogonal group.

Let M be a smooth manifold of (real) dimension $m := 2\bar{m}$. We say that J is an *integrable complex structure* on the tangent bundle TM if there is a coordinate atlas with local coordinates $(x^1, \dots, x^m)$ so that

$$J\partial_{x^i} = \partial_{x^{i+\bar{m}}} \quad \text{and} \quad J\partial_{x^{i+\bar{m}}} = -\partial_{x^i} \quad \text{for} \quad 1 \le i \le \bar{m}\,. \tag{9.4.a}$$

We say that a pseudo-Riemannian metric h on TM is *pseudo-Hermitian* if $J^*h = h$. Let ∇^h be the Levi–Civita connection of h. Since ∇^h need not commute with J, we average over the action of J to define a J-invariant connection by setting:

$$\tilde{\nabla}^h := \tfrac{1}{2}\{\nabla^h + J^*\nabla^h\} = \tfrac{1}{2}\{\nabla^h - J\nabla^h J\}\,.$$

Let R_c^h be the associated complex curvature tensor. If $\Theta \in \mathfrak{C}_{k,\ell}$, then we may form

$$\Theta(h) := \Theta(R_c^h) \in \Lambda^{2k}(M) \otimes_{\mathbb{R}} \mathbb{C}\,.$$

Of particular interest is the special case where the Levi–Civita connection actually does commute with J, i.e., $\nabla^h J = 0$ and in that setting, the triple (M, h, J) is said to be a *pseudo-Kähler* manifold. If h is positive definite, then (M, h, J) is said to be a *Kähler* manifold. The pseudo-Kähler condition is very strong. One feature is that there exist normal-holomorphic coordinates, i.e., holomorphic coordinate systems where the first derivatives of the metric vanish. If (M, h, J) is a pseudo-Kähler manifold, then $\tilde{\nabla}^h = \nabla^h$. Let $c_\ell(A) := \det(\frac{\sqrt{-1}}{2\pi}A)$. We then have that $c_\ell \in \mathfrak{C}_{\ell,\ell}$ and $c_\ell(h) = E_{2\ell}\,dv_g$ is the integrand of the Chern–Gauss–Bonnet Theorem.

Let $\Omega_h(x, y) := h(x, Jy)$. We assume (M, h, J) is pseudo-Kähler; this implies that $d\Omega_h = 0$. Any complex manifold inherits a natural orientation so we may identify measures with m-forms. In the positive definite setting, we have under this identification that $dv_g = \frac{1}{\bar{m}!}\Omega_h^{\bar{m}}$ and we replace dv_g by $\frac{1}{\bar{m}!}\Omega_h^{\bar{m}}$ to avoid complications with signs henceforth in the higher signature context.

Let $\Theta \in \mathfrak{C}_{k,\ell}$. As noted above, $\Theta(h)$ is a differential form of degree $2k$. To obtain a scalar invariant, we contract with the Kähler form $\Omega_h(x, y) := h(x, Jy)$ and in analogy with Gauss–Bonnet gravity (as discussed in Section 9.3.1), we consider the scalar invariant $h(\Theta(h), \Omega_h^k)$ and associated Lovelock functional

$$\Theta[M,h,J] := \tfrac{1}{\bar{m}!}\int_M h(\Theta(h),\Omega_h^k)\cdot\Omega_h^{\bar{m}}\,.$$

The associated Euler–Lagrange equations are defined by setting:

$$\mathrm{EL}_\Theta(h,\kappa) := \tfrac{1}{\bar{m}!}\left\{\partial_\epsilon\int_M h(\Theta(h+\epsilon\kappa),\Omega_{h+\epsilon k}^k)\cdot\Omega_{h+\epsilon\kappa}^{\bar{m}}\right\}\Big|_{\epsilon=0},$$

where κ is a J-invariant symmetric 2-tensor with compact support. If $\bar{m}=k$ and if M is pseudo-Kähler, then $\Theta[M,h,J]$ is independent of h and, consequently, the associated Euler–Lagrange equations vanish. We therefore suppose that $k<\bar{m}$. We can integrate by parts to express

$$\mathrm{EL}_\Theta(h,\kappa) = \tfrac{1}{\bar{m}!}\int_M \langle\mathcal{E}_\Theta(h),\kappa\rangle\cdot\Omega_h^{\bar{m}},$$

where $\mathcal{E}_\Theta(h)\in S^2(TM)$ is a J-invariant symmetric 2-tensor field and where $\langle\cdot,\cdot\rangle$ denotes the natural pairing between $S^2(TM)$ and $S^2(T^*M)$. Let

$$\mathcal{F}_\Theta(h) := -\tfrac{1}{k+1}\sqrt{-1}h(\Theta(R_c^h)\wedge e^\alpha\wedge e^{\bar{\beta}})\otimes e_\alpha\otimes e_{\bar{\beta}}\,.$$

Of course, for a general h, $\mathcal{E}_\Theta(h)$ is very complicated and cannot be expressed directly in terms of curvature. However, after a lengthy and difficult calculation in invariance theory, it was shown in Gilkey, Park and Sekigawa [78] that

Theorem 9.42 *Let $\Theta\in\mathfrak{C}_{k,\bar{m}}$. If (M,h,J) is a Kähler manifold, $\mathcal{E}_\Theta(h)=\mathcal{F}_\Theta(h)$.*

The main result of this section extends Theorem 9.42 to the pseudo-Kähler setting.

Theorem 9.43 *Let $\Theta\in\mathfrak{C}_{k,\bar{m}}$. If (M,h,J) is a pseudo-Kähler manifold, $\mathcal{E}_\Theta(h)=\mathcal{F}_\Theta(h)$.*

One could redo the analysis of Gilkey, Park and Sekigawa [78] taking into account the fact that the structure group $U(p,q)$ involves not only rotations but also hyperbolic boosts. But instead, we will use analytic continuation to pass from the positive definite to the indefinite setting.

Let (M,h,J) be a pseudo-Kähler manifold of signature $(2\bar{p},2\bar{q})$. Fix a point P of M and choose local coordinates $\vec{x}=(x^1,\dots,x^m)$ centered at P so that J is given by Equation (9.4.a). We complexify and set, for $1\le\alpha\le\bar{m}$,

$$\begin{aligned}\partial_{z^\alpha} &:= \tfrac{1}{2}\{\partial_{x^\alpha}-\sqrt{-1}\partial_{x^{\bar{m}+\alpha}}\}, & dz^\alpha &:= \{dx^\alpha+\sqrt{-1}dx^{\bar{m}+\alpha}\},\\ \partial_{\bar{z}^\alpha} &:= \tfrac{1}{2}\{\partial_{x^\alpha}+\sqrt{-1}\partial_{x^{\bar{m}+\alpha}}\}, & d\bar{z}^\alpha &:= \{dx^\alpha-\sqrt{-1}dx^{\bar{m}+\alpha}\}.\end{aligned}$$

We extend the metric h to be complex bilinear and set $h_{\alpha,\bar\beta} := h(\partial_{z^\alpha}, \partial_{\bar z^\beta})$. The condition that $J^*h = h$ is then equivalent to the identities:

$$h(\partial_{z^\alpha}, \partial_{z^\beta}) = 0, \quad h(\partial_{\bar z^\alpha}, \partial_{\bar z^\beta}) = 0, \quad \bar h(\partial_{z^\alpha}, \partial_{\bar z^\beta}) = h(\partial_{z^\beta}, \partial_{\bar z^\alpha}).$$

We set $h_{\alpha,\bar\beta} := h(\partial_{z^\alpha}, \partial_{\bar z^\beta})$. We then have $\bar h_{\alpha,\bar\beta} = h_{\beta,\bar\alpha}$. If we set $h_{\alpha,\bar\beta/\gamma} := \partial_{z^\gamma} h_{\alpha,\bar\beta}$ and $h_{\alpha,\bar\beta/\bar\gamma} := \partial_{\bar z^\gamma} h_{\alpha,\bar\beta}$, the Kähler condition becomes

$$h_{\alpha,\bar\beta/\gamma} = h_{\gamma,\bar\beta/\alpha} \quad \text{and} \quad h_{\alpha,\bar\beta/\bar\gamma} = h_{\alpha,\bar\gamma/\bar\beta}. \tag{9.4.b}$$

Let $A := (\alpha_1, \dots, \alpha_\nu)$ and $B := (\beta_1, \dots, \beta_\mu)$ be collection of indices between 1 and $\bar m$. Set

$$h(A; B) := \partial_{z^{\alpha_2}} \cdots \partial_{z^{\alpha_\nu}} \partial_{\bar z^{\beta_2}} \cdots \partial_{\bar z^{\beta_\mu}} h_{\alpha_1,\bar\beta_1}.$$

It is immediate that $\bar h(A; B) = h(B; A)$. Differentiate Equation (9.4.b) to see that one can permute the elements of A and the elements of B without changing $h(A; B)$. The following Lemma was proved in Gilkey, Park and Sekigawa [78] in the positive definite setting. The proof, however, involved quadratic and higher-order holomorphic changes and was independent of the signature of the metric. It extends without change to the setting at hand.

Lemma 9.44 Let P be a point of a Kähler manifold (M, J, h) of dimension $m = 2\bar m$. Fix $n \ge 2$. There exist local holomorphic coordinates $\vec x = (x^1, \dots, x^{2\bar m})$ centered at P so that

1. J is given by Equation (9.4.a).
2. $h(A; B)(P) = 0$ for $|B| = 1$ and $2 \le |A| \le n$.

Suppose given complex constants $c(A; B)$ for $2 \le |A| \le n$ and $2 \le |B| \le n$ which are such that $c(A; B) = \bar c(B; A)$. There exists a Kähler metric $\tilde h$ on (M, J) so that $\tilde h(A; B) = 0$ for $|B| = 1$ and so that $\tilde h(A; B)(P) = c(A; B)$ for $2 \le |A| \le n$ and $2 \le |B| \le n$.

The variables $\{h(A; B)\}$ are a good choice of variables since, unlike the covariant derivatives of the curvature tensor, there are no additional identities and we are dealing with a pure polynomial algebra.

Proof. We now establish Theorem 9.43. Fix $\Theta \in \mathfrak{C}_{k,\bar m}$. We work purely formally. We use Lemma 9.44 to regard $\mathcal{E}_\Theta$ and $\mathcal{F}_\Theta$ as polynomials in $h(P)$, the variables $h(A; B)$, and $\det(h)^{-1}$ where h is a J-invariant symmetric bilinear form on $\mathbb{R}^m$; we must introduce the variable $\det(h)^{-1}$ to define the metric on the cotangent bundle and to raise and lower indices. These polynomials are well-defined if $\det(h) \ne 0$ and we have $\mathcal{E}_\Theta - \mathcal{F}_\Theta = 0$ if h is positive definite. But, of course, we can allow h to be complex-valued. We have that $\mathcal{E}_\Theta - \mathcal{F}_\Theta = 0$ if h is real-valued and positive definite. Imposing the condition $\det(h) \ne 0$ does not disconnect the parameter space. Consequently, the identity theorem yields $\mathcal{E}_\Theta - \mathcal{F}_\Theta = 0$ in complete generality and, in particular, if h has indefinite signature. $\square$

9.5 VSI MANIFOLDS

A pseudo-Riemannian manifold is said to be a *VSI manifold* (vanishing scalar invariants) if all the Weyl scalar invariants vanish. If M is not flat, this implies that the metric in question has indefinite signature since $\|R\|^2$ is a scalar invariant which vanishes in the Riemannian setting if and only if M is flat. We refer to the discussion in Alcolado et al. [2], Coley et al. [50], and Coley, Hervik and Pelavas [48, 49] for further details.

Definition 9.45 Let $\vec{x} = (x^1, \dots, x^m)$ be the usual coordinates on $M = \mathbb{R}^m$. An affine manifold $\mathcal{M} := (\mathbb{R}^m, \nabla)$ is said to be a *generalized plane wave manifold* if

$$\nabla_{\partial_{x^i}} \partial_{x^j} = \sum_{k>\max(i,j)} \Gamma_{ij}{}^k(x^1, \dots, x^{k-1}) \partial_{x^k} . \tag{9.5.a}$$

If the connection on a generalized plane wave manifold arises from the Levi–Civita connection of a pseudo-Riemannian metric on M, then the manifold is VSI. We say that an affine manifold (M, ∇) is *geodesically convex* if between any two points of M there is a unique geodesic. The following result applies to the manifolds described in Example 9.18 and was originally motivated by these examples:

Theorem 9.46 *Let $\mathcal{M}$ be a generalized plane wave manifold.*

1. *$\mathcal{M}$ is geodesically complete and geodesically convex.*
2. *The exponential map $\exp_P$ is a diffeomorphism from $T_P M$ to M for any point $P \in M$.*
3. $\nabla_{\partial_{j_1}} \cdots \nabla_{\partial_{j_\nu}} \mathcal{R}(\partial_{i_1}, \partial_{i_2})\partial_{i_3} = \sum_{k>\max(i_1,i_2,i_3,j_1,\dots,j_\nu)} \mathcal{R}_{i_1 i_2 i_3}{}^k{}_{;j_1 \dots j_\nu}(x^1, \dots, x^{k-1})\partial_k$.
4. *If ∇ is the Levi–Civita connection of a metric g on M, then M is VSI.*

Proof. It is rather rare in affine geometry to be able to solve the geodesic equations explicitly. In this instance, however, we can use a recursive formalism. We must solve the equation:

$$\begin{array}{l} \ddot{\gamma}^k(t) + \sum_{i,j<k} \dot{\gamma}^i(t)\dot{\gamma}^j(t)\Gamma_{ij}{}^k(\gamma^1, \dots, \gamma^{k-1})(t) = 0 \\ \text{with} \quad \gamma^k(0) = \gamma_0^k \quad \text{and} \quad \dot{\gamma}^k(0) = \gamma_1^k . \end{array} \tag{9.5.b}$$

Set $\gamma^1(t) := \gamma_0^1 + \gamma_1^1 t$. For $k > 1$, we proceed inductively to solve Equation (9.5.b) by setting:

$$\gamma^k(t) := \gamma_0^k + \gamma_1^k t - \int_0^t \int_0^s \sum_{i,j<k} \dot{\gamma}^i(r)\dot{\gamma}^j(r)\Gamma_{ij}{}^k(\gamma^1, \dots, \gamma^{k-1})(r) dr ds .$$

Thus, every geodesic arises in this way so all geodesics extend for infinite time. Furthermore, given $P, Q \in M$, there is a unique geodesic $\gamma = \gamma_{P,Q}$ so that $\gamma(0) = P$ and $\gamma(1) = Q$ where $\gamma_0^k = P^k$, $\gamma_1^1 = Q_1 - P_1$ and for $k > 1$,

$$\gamma_1^k = Q^k - P^k + \int_0^1 \int_0^s \sum_{i,j<k} \dot{\gamma}^i(r)\dot{\gamma}^j(r)\Gamma_{ij}{}^k(\gamma^1, \dots, \gamma^{k-1})(r)drds\,.$$

Assertion 1 and Assertion 2 now follow. To prove Assertion 3, we expand

$$\begin{aligned} R_{ijk}{}^\ell \quad = \quad & \partial_{x^i}\Gamma_{jk}{}^\ell(x^1, \dots, x^{\ell-1}) - \partial_{x^j}\Gamma_{ik}{}^\ell(x^1, \dots, x^{\ell-1}) \\ & +\Gamma_{in}{}^\ell(x^1, \dots, x^{\ell-1})\Gamma_{jk}{}^n(x^1, \dots, x^{n-1}) \\ & -\Gamma_{jn}{}^\ell(x^1, \dots, x^{\ell-1})\Gamma_{ik}{}^n(x^1, \dots, x^{n-1})\,. \end{aligned}$$

The sums involving the quadratic terms range over $n < \ell$ so $R_{ijk}{}^\ell = R_{ijk}{}^\ell(x^1, \dots, x^{\ell-1})$. Suppose $\ell \le k$. Then $\Gamma_{jk}{}^\ell = \Gamma_{ik}{}^\ell = 0$. If either of the quadratic terms is non-zero, then there exists an index n with $k < n$ and $n < \ell$. This is not possible if $\ell \le k$. Thus, $R_{ijk}{}^\ell = 0$ if $\ell \le k$. If $\ell \le i$, then $\partial_{x^i}\Gamma_{jk}{}^\ell(x^1, \dots, x^{\ell-1}) = 0$ and $\partial_{x^j}\Gamma_{ik}{}^\ell = \partial_{x^j}0 = 0$. We have $\Gamma_{in}{}^\ell = 0$. For the other quadratic term to be non-zero, there must exist an index n so $i < n$ and $n < \ell$. This is not possible if $\ell \le i$. This shows $R_{ijk}{}^\ell = 0$ if $\ell \le i$; similarly $R_{ijk}{}^\ell = 0$ if $\ell \le j$. Assertion 3 follows if $\nu = 0$. To study ∇R, we expand

$$\begin{aligned} R_{ijk}{}^n{}_{;\ell} \quad = \quad & \partial_\ell R_{ijk}{}^n(x^1, \dots, x^{n-1}) && (9.5.c) \\ & -R_{rjk}{}^n(x^1, \dots, x^{n-1})\Gamma_{\ell i}{}^r(x^1, \dots, x^{r-1}) && (9.5.d) \\ & -R_{irk}{}^n(x^1, \dots, x^{n-1})\Gamma_{\ell j}{}^r(x^1, \dots, x^{r-1}) && (9.5.e) \\ & -R_{ijr}{}^n(x^1, \dots, x^{n-1})\Gamma_{\ell k}{}^r(x^1, \dots, x^{r-1}) && (9.5.f) \\ & -R_{ijk}{}^r(x^1, \dots, x^{r-1})\Gamma_{\ell r}{}^n(x^1, \dots, x^{n-1})\,. && (9.5.g) \end{aligned}$$

We have $r < n$ in these equations. Thus, $R_{ijk}{}^n{}_{;\ell} = R_{ijk}{}^n{}_{;\ell}(x^1, \dots, x^{n-1})$. Let $\nu = 1$. The following steps show that $R_{ijk}{}^n{}_{;\ell} = 0$ if $n \le \max(i, j, k, \ell)$.

1. $\partial_\ell R_{ijk}{}^n(x^1, \dots, x^{n-1}) = 0$ if $n \le \max(i, j, k, \ell)$ in Equation (9.5.c).
2. $n > \max(r, j, k)$ and $r > \max(i, \ell)$ so $n > \max(i, j, k, \ell)$ in Equation (9.5.d).
3. $n > \max(i, r, k)$ and $r > \max(\ell, j)$ so $n > \max(i, j, k, \ell)$ in Equation (9.5.e).
4. $n > \max(i, j, r)$ and $r > \max(k, \ell)$ so $n > \max(i, j, k, \ell)$ in Equation (9.5.f).
5. $n > \max(\ell, r)$ and $r > \max(i, j, k)$ so $n > \max(i, j, k, \ell)$ in Equation (9.5.g).

This establishes Assertion 3 if $\nu = 1$ so we are dealing with ∇R. The argument is the same for higher values of ν and is therefore omitted.

We now turn to the question of Weyl invariants. Suppose ∇ is the Levi–Civita connection of a pseudo-Riemannian metric on M. Let Θ be a Weyl monomial which is formed by contracting upper and lower indices in pairs in the variables $\{g^{ij}, g_{ij}, R_{i_1 i_2 i_3}{}^{i_4}{}_{;j_1 \dots}\}$. The single upper index in R plays a distinguished role. We choose a representation for Θ so the number of g_{ij} variables is minimal. Suppose there is a g_{ij} variable in this minimal representation; this means that $\Theta = g_{ij} R_{u_1u_2u_3}{}^{i}{}_{;\dots} R_{v_1v_2v_3}{}^{j}{}_{;\dots} \cdots$. Suppose further that $g^{u_1w_1}$ appears in Θ; consequently, $\Theta = g_{ij} g^{u_1w_1} R_{u_1u_2u_3}{}^{i}{}_{;\dots} R_{v_1v_2v_3}{}^{j}{}_{;\dots} \cdots$. We could then raise and lower an index to express

$$\Theta = R^{w_1}{}_{u_2u_3j;\dots} R_{v_1v_2v_3}{}^{j}{}_{;\dots} \cdots = R_{ju_3u_2}{}^{w_1}{}_{;\dots} R_{v_1v_2v_3}{}^{j}{}_{;\dots} \cdots .$$

This has one less g_{**} variable. This contradicts the assumed minimality. Thus, u_1 must be contracted against an upper index; a similar argument shows that u_2, u_3, v_1, v_2, and v_3 are contracted against an upper index as well. Consequently,

$$\Theta = g_{ij} R_{u_1u_2u_3}{}^{i}{}_{;\dots} R_{v_1v_2v_3}{}^{j}{}_{;\dots} R_{w_1w_2w_3}{}^{u_1}{}_{;\dots} \cdots .$$

Suppose w_1 is not contracted against an upper index. We then have

$$\Theta = g_{ij} g^{w_1x_1} R_{u_1u_2u_3}{}^{i}{}_{;\dots} R_{v_1v_2v_3}{}^{j}{}_{;\dots} R_{w_1w_2w_3}{}^{u_1}{}_{;\dots} \cdots ,$$

where we use the curvature symmetries, the covariant derivative of the second Bianchi identity and, if necessary, commute covariant derivatives at the cost of introducing additional curvature terms to ensure that the index w_1 appears in the position indicated. Thus,

$$\begin{aligned} \Theta &= R_{u_1u_2u_3j;\dots} R_{v_1v_2v_3}{}^{j}{}_{;\dots} R^{x_1}{}_{w_2w_3}{}^{u_1}{}_{;\dots} \cdots \\ &= g^{u_1y_1} R_{u_1u_2u_3j;\dots} R_{v_1v_2v_3}{}^{j}{}_{;\dots} R^{x_1}{}_{w_2w_3y_1;\dots} \cdots \\ &= R^{y_1}{}_{u_2u_3j;\dots} R_{v_1v_2v_3}{}^{j}{}_{;\dots} R^{x_1}{}_{w_2w_3y_1;\dots} \\ &= R_{ju_3u_2}{}^{y_1}{}_{;\dots} R_{v_1v_2v_3}{}^{j}{}_{;\dots} R_{w_2w_3y_1}{}^{x_1}{}_{;\dots} \end{aligned}$$

which has one less g_{ij} variable. Thus, w_1 is contracted against an upper index so

$$\Theta = g_{ij} R_{u_1u_2u_3}{}^{i}{}_{;\dots} R_{v_1v_2v_3}{}^{j}{}_{;\dots} R_{w_1w_2w_3}{}^{u_1}{}_{;\dots} R_{x_1x_2x_3}{}^{w_1}{}_{;\dots} \cdots .$$

We continue in this fashion to build a monomial of infinite length. This is not possible. Thus, we can always find a representation for Θ which contains no g_{ij} variables in the summation. We suppose the evaluation of Θ is non-zero and argue for a contradiction. To simplify the notation, group all the lower indices together. By considering the pairing of upper and lower indices, we see that we can expand Θ in cycles $\Theta = R_{\dots i_r \dots}{}^{i_1} R_{\dots i_1 \dots}{}^{i_2} \cdots R_{\dots i_{r-1} \dots}{}^{i_r} \cdots$. By Assertion 3, $R_{\dots j \dots}{}^{\ell} = 0$ if $\ell \le j$ so the sum runs over indices $i_r < i_1 < i_2 < \cdots < i_r$. As this is the empty sum, we see that $\Theta = 0$ as desired and the final assertion follows. □

Not every manifold that is VSI is a generalized plane wave manifold. Let $\{x, y, \tilde{x}\}$ be coordinates on $\mathbb{R}^3$. If $f = f(x, y)$ is a smooth function of 2-variables, let

$$g_f(\partial_x, \partial_x) := -2f(x, y), \quad g_f(\partial_x, \partial_{\tilde{x}}) := 1, \quad g_f(\partial_y, \partial_y) := 1 .$$

We will show in Lemma 10.20 that any such manifold is VSI. If we take $f(x, y) = y^2$, then we obtain a symmetric space. But if we take $f(x, y) = -y^n$ for $n = 3, 4, \dots$, $\mathcal{M}_f$ is geodesically incomplete and exhibits Ricci blowup; such a manifold cannot be embedded in a geodesically complete manifold.

We continue our discussion of the manifolds in Example 9.18.

Lemma 9.47 The manifolds of Example 9.18 are generalized plane wave manifolds. Thus, they are VSI and geodesically complete.

Proof. We recall the definition of these manifolds. Let $F := \{f(y) + yz^0 + \cdots + y^{\ell+1}z^\ell\}$. Let $\vec{u} = (u^0, \dots, v^{\ell+2}) = (x, y, z^0, \dots, z^\ell)$ and $\vec{v} := (v^0, \dots, v^{\ell+2}) = (\tilde{x}, \tilde{y}, \tilde{z}^0, \dots, \tilde{z}^\ell)$. Then

$$g_f(\partial_{u^0}, \partial_{u^0}) := -2F \quad \text{and} \quad g_f(\partial_{u^i}, \partial_{\tilde{u}^i}) := 1 \quad \text{for} \quad 0 \leq i \leq \ell + 2\,.$$

The (possibly) non-zero covariant derivatives are given by:

$$\begin{aligned} &g(\nabla_{\partial_{u^0}} \partial_{u^0}, \partial_{u^i}) = \partial_{u^i} F \quad \text{for} \quad i \geq 1, \quad g(\nabla_{\partial_{u^0}} \partial_{u^0}, \partial_{u^0}) = 0, \\ &g(\nabla_{\partial_{u^0}} \partial_{u^i}, \partial_{u^0}) = g(\nabla_{\partial_{u^i}} \partial_{u^0}, \partial_{u^i}) = \partial_{u^i} F \quad \text{for} \quad i \geq 1\,. \end{aligned}$$

From these equations and from the form of the metric, we conclude the (possibly) non-zero covariant derivatives are:

$$\nabla_{\partial^i_u} \partial_{u^j} = C_{ij}{}^k(u) \partial_{\tilde{u}^k}\,.$$

This has the form given in Equation (9.5.a). □

9.6 INVARIANTS THAT ARE NOT OF WEYL TYPE

Invariants which are not of Weyl type frequently arise in affine geometry since there is often, but not always, no auxiliary non-singular symmetric bilinear form to use in contracting indices. To illustrate this point, we discuss, briefly, the classification of homogeneous surfaces of Brozos-Vázquez, García-Río and Gilkey [19, 20]. In Section 9.6.1, we review the theory of homogeneous affine surfaces. In Section 9.6.2, we present an example of an invariant which is not of Weyl type. In Section 9.6.3, we present an example where the invariant is of Weyl type where the metric is given by the associated Ricci tensor. In Section 9.6.4, we discuss k-curvature homogeneity.

9.6.1 HOMOGENEOUS AFFINE SURFACES. An affine manifold $\mathcal{M} = (M, \nabla)$ is said to be *locally homogeneous* if given any two points of M, there is the germ of a diffeomorphism Φ taking one point to another with $\Phi^*\nabla = \nabla$. One has the following classification of locally homogeneous affine surfaces which is due to Opozda [111] (see Arias-Marco and Kowalski [7] for an extension allowing non-zero torsion):

Barbara Opozda

Theorem 9.48 *Let $\mathcal{M} = (M, \nabla)$ be a locally homogeneous affine surface. Then at least one of the following three possibilities holds which describe the local geometry.*

1. *There exist local coordinates (x^1, x^2) so that $\Gamma_{ij}{}^k = \Gamma_{ji}{}^k \in \mathbb{R}$.*
2. *There exist local coordinates (x^1, x^2) so that $\Gamma_{ij}{}^k = (x^1)^{-1} C_{ij}{}^k$ for $C_{ij}{}^k = C_{ji}{}^k \in \mathbb{R}$.*
3. *∇ is the Levi–Civita connection of a metric of constant sectional curvature.*

An affine surface $\mathcal{M}$ is said to be *Type $\mathcal{A}$* (resp. *Type $\mathcal{B}$ or Type $\mathcal{C}$*) if $\mathcal{M}$ is locally homogeneous, if $\mathcal{M}$ is not flat, and if Assertion 1 (resp. Assertion 2 or Assertion 3) of Theorem 9.48 holds. There are no surfaces which are both Type $\mathcal{A}$ and Type $\mathcal{C}$. But there are surfaces which are both Type $\mathcal{A}$ and Type $\mathcal{B}$ and there are surfaces which are both Type $\mathcal{B}$ and Type $\mathcal{C}$. We will return to these geometries when we discuss affine Killing vector fields on affine surfaces in Theorem 11.43 and Theorem 11.44.

Recall that a vector field X is said to be a *Killing vector field* if $\mathcal{L}_X g = 0$; for further details we refer to Lemma 7.8 of Book II. There is a similar notion in the affine setting we present below; we will postpone a discussion of homothety vector fields until we establish Lemma 10.5. We refer to Kobayashi and Nomizu [85, 86] for the proof of the following result; the affine Killing vector fields play the same role in affine geometry that Killing vector fields play in pseudo-Riemannian geometry.

Lemma 9.49 If X is a smooth vector field on M, let Φ_t^X be the local flow defined by X. The following conditions are equivalent and if any is satisfied, we will say that X is an *affine Killing vector field.*

1. $(\Phi_t^X)_* \circ \nabla = \nabla \circ (\Phi_t^X)_*$ on the appropriate domain.
2. The Lie derivative $\mathcal{L}_X(\nabla)$ of ∇ vanishes.
3. $[X, \nabla_Y Z] - \nabla_Y [X, Z] - \nabla_{[X,Y]} Z = 0$ for all $Y, Z \in C^\infty(TM)$.

Let $\mathfrak{K}(\mathcal{M})$ be the set of affine Killing vector fields. The Lie bracket gives $\mathfrak{K}(\mathcal{M})$ the structure of a real Lie algebra. Furthermore, if $X \in \mathfrak{K}(\mathcal{M})$, if $X(P) = 0$, and if $\nabla X(P) = 0$, then $X \equiv 0$. Let $\kappa(\mathcal{M}) = \dim(\mathfrak{K}(\mathcal{M}))$.

Let $\mathcal{M}$ be a Type $\mathcal{A}$ surface. The Ricci tensor of $\mathcal{M}$ is symmetric; this is not always the case in affine geometry and in particular there are Type $\mathcal{B}$ surfaces where the Ricci tensor is alternating and non-zero. If the Ricci tensor has rank 2, then $\kappa(\mathcal{M}) = 2$ while if the Ricci tensor has rank 1, then $\kappa(\mathcal{M}) = 4$. Thus, the rank of the Ricci tensor determines much of the geometry of $\mathcal{M}$ in the Type $\mathcal{A}$ setting.

9.6.2 TYPE $\mathcal{A}$ SURFACES WITH RANK$(\rho) = 1$. Suppose Rank$(\rho) = 1$. Choose a tangent vector $X \in T_P M$ so $\rho(X, X) \neq 0$ and set

$$\alpha_X(\mathcal{M}) := \nabla\rho(X, X; X)^2 \cdot \rho(X, X)^{-3} \quad \text{and} \quad \epsilon_X(\mathcal{M}) := \operatorname{Sign}\{\rho(X, X)\} = \pm 1\,.$$

The following result is established in Brozos-Vázquez, García-Río and Gilkey [20].

Theorem 9.50 *Let $\mathcal{M}$ and $\tilde{\mathcal{M}}$ be Type $\mathcal{A}$ surfaces with* Rank$(\rho) = 1$.

1. *There exists a 1-form ω so that $\nabla^k \rho = (k+1)!\omega^k \otimes \rho$ for any k.*
2. *$\alpha(\mathcal{M}) := \alpha_X(\mathcal{M})$ and $\epsilon(\mathcal{M}) := \epsilon_X(\mathcal{M})$ are independent of X.*
3. *$\mathcal{M}$ is affine equivalent to $\tilde{\mathcal{M}}$ if and only if $\alpha(\mathcal{M}) = \alpha(\tilde{\mathcal{M}})$ and $\epsilon(\mathcal{M}) = \epsilon(\tilde{\mathcal{M}})$.*

The invariant α takes values in $\mathbb{R}$ and the invariant ϵ takes values in $\mathbb{Z}_2$. One can use these two invariants to show that the moduli space of Type $\mathcal{A}$ structures with Rank$(\rho) = 1$ consists of two disjoint components one of which is isomorphic (via α) to $(-\infty, 0]$ and the other of which is isomorphic (via α) to $[0, \infty)$. Thus, the invariant α plays a crucial role in the analysis but it is definitely not of Weyl type.

9.6.3 TYPE $\mathcal{A}$ SURFACES WITH RANK$(\rho) = 2$. Suppose, on the other hand, that Rank$(\rho) = 2$. The Ricci tensor then defines an auxiliary pseudo-Riemannian metric on the locally homogeneous affine surface $\mathcal{M}$. Since the components of ρ are constant, ρ is a flat metric. It then follows that the Type $\mathcal{A}$ coordinate atlas on $\mathcal{M}$ is given by affine linear transformations of the form $(x^1, x^2) \to (a_1^1 x^1 + a_2^1 x^2 + b^1, a_1^2 x^1 + a_2^2 x^2 + b^2)$ where (a_i^j) belongs to $\mathrm{GL}(2, \mathbb{R})$. Sum over repeated indices to define $\rho_{ij}^3 := \Gamma_{ik}{}^\ell \Gamma_{j\ell}{}^k$. This is then a symmetric 2-tensor. Since contracting a lower index against an upper index is an affine invariant, ρ^3 is invariantly defined on $\mathcal{M}$. Define

$$\psi_3 := \operatorname{Tr}_\rho\{\rho^3\} = \rho^{ij}\rho_{ij}^3 \quad \text{and} \quad \Psi_3 := \det(\rho^3)/\det(\rho)\,.$$

These are then scalar invariants in this setting which are of Weyl type despite the fact that we are working in affine geometry. Let $\mathcal{Z}_+$ (resp. $\mathcal{Z}_0$ or $\mathcal{Z}_-$) be the set of Christoffel symbols $\Gamma \in \mathbb{R}^6$ defining a Type $\mathcal{A}$ structure such that the Ricci tensor is positive (resp. indefinite or negative) definite. Let $\mathfrak{Z}_\varepsilon$ for $\varepsilon = +, 0, -$ be the associated moduli space. Let $\Theta_\varepsilon := (\psi_3, \Psi_3)$ on $\mathcal{Z}_\varepsilon$; this real analytic map extends to a map from the moduli space $\mathfrak{Z}_\varepsilon$ to $\mathbb{R}^2$. Consider the curves $\sigma_\pm(t) := (\pm 4t^2 \pm \frac{1}{t^2} + 2, 4t^4 \pm 4t^2 + 2)$. The curve σ_+ is smooth; the curve σ_- has a cusp at $(-2, 1)$ when $t = \frac{1}{\sqrt{2}}$; it corresponds to the structure

$$\Gamma_{11}{}^1 = 1, \quad \Gamma_{11}{}^2 = 0, \quad \Gamma_{12}{}^1 = 0, \quad \Gamma_{12}{}^2 = -1, \quad \Gamma_{22}{}^1 = -1, \quad \Gamma_{22}{}^2 = 0\,.$$

These two curves divide the plane into three open regions $\mathfrak{D}_-$, $\mathfrak{D}_0$, and $\mathfrak{D}_+$ where $\mathfrak{D}_-$ lies in the second quadrant and is bounded on the right by σ_-, $\mathfrak{D}_+$ lies in the first quadrant, and is bounded on the left by σ_+ and $\mathfrak{D}_0$ lies in between and is bounded on the left by σ_- and on the right by σ_+. Let $\mathfrak{C}_\varepsilon$ be the closure of $\mathfrak{D}_\varepsilon$;

$$\mathfrak{C}_- = \mathfrak{D}_- \cup \operatorname{range}\{\sigma_-\}, \quad \mathfrak{C}_+ = \mathfrak{D}_+ \cup \operatorname{range}\{\sigma_+\}, \quad \mathfrak{C}_0 = \mathfrak{D}_0 \cup \operatorname{range}\{\sigma_-\} \cup \operatorname{range}\{\sigma_+\}.$$

We refer to Brozos-Vázquez, García-Río and Gilkey [19] for the proof of the following result.

Theorem 9.51 *Θ_ε is 1-1 map from $\mathfrak{Z}_\varepsilon$ to $\mathfrak{C}_\varepsilon$.*

Next, we give the two curves $\sigma_\pm$ which bound the moduli spaces and also two different pictures of the moduli spaces:

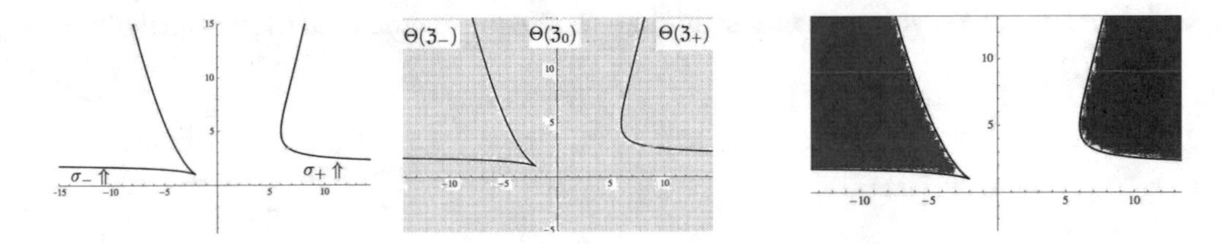

Θ_ε is 1-1 on $\mathfrak{Z}_\varepsilon$, $\Theta_+(\mathfrak{Z}_+)$ intersects $\Theta_0(\mathfrak{Z}_0)$ along their common boundary σ_+ and $\Theta_-(\mathfrak{Z}_+)$ intersects $\Theta_0(\mathfrak{Z}_0)$ along their common boundary σ_-. This does not mean that $\mathfrak{Z}_+$ intersects $\mathfrak{Z}_0$, nor does it mean that $\mathfrak{Z}_1$ intersects $\mathfrak{Z}_0$, nor does it mean that Θ_0 or Θ_+ or $\Theta_-$are not 1-1 on their respective domains.

9.6.4 K-CURVATURE HOMOGENEITY. We follow the discussion in Gilkey [73, Cha. 2] for the material of this section. Let $\mathcal{M} = (M, \nabla)$ be an affine manifold. Fix an integer $\ell \geq 0$. Let $(x, y, z^0, \ldots, z^\ell, \tilde{x}, \tilde{y}, \tilde{z}^0, \ldots, \tilde{z}^\ell)$ be the usual coordinates on $\mathbb{R}^{6+2\ell}$. Let $f \in C^\infty(\mathbb{R})$ satisfy $f^{(\ell+3)} > 0$ and $f^{(\ell+4)} > 0$. Let

$$F(y, \vec{z}) := f(y) + yz^0 + \cdots + y^{\ell+1} z^\ell .$$

Let $\mathcal{M}_f := (\mathbb{R}^{6+2\ell}, g)$ be the pseudo-Riemannian manifold of neutral signature $(\ell+3, \ell+3)$ where $g(\partial_x, \partial_{\tilde{x}}) = g(\partial_y, \partial_{\tilde{y}}) = g(\partial_{z^i}, \partial_{\tilde{z}^i}) = 1$ and $g(\partial_x, \partial_x) = -2F(y, \vec{z})$. For $\mu \geq 3$, let

$$\alpha_\mu := f^{(\ell+\mu+2)}(f^{(\ell+3)})^{\mu-2}(f^{(\ell+4)})^{1-\mu} .$$

The proof of the following result illustrates the use of affine invariants α_μ which are not of Weyl type.

Theorem 9.52

1. *$\mathcal{M}_f$ is a generalized plane wave manifold.*
2. *α_μ is an affine invariant.*
3. *If f_1 and f_2 are real analytic and if $\alpha_\mu(f_1)(P_1) = \alpha_\mu(f_2)(P_2)$ for all $\mu \geq 3$, then there is an isometry $\Phi : \mathcal{M}_{f_1} \to \mathcal{M}_{f_2}$ with $\Phi(P_1) = P_2$.*
4. *The following assertions are equivalent.*
   (a) *$\mathcal{M}$ is $(\ell + 3)$-affine curvature homogeneous.*
   (b) *α_3 is constant.*
   (c) *$f^{(\ell+3)}(y) = ae^{by}$ for some $a > 0$ and $b > 0$.*

d. *$\mathcal{M}$ is a homogeneous pseudo-Riemannian manifold.*

Since these pseudo-Riemannian manifolds are generalized plane wave manifolds, Theorem 9.46 shows that all the scalar Weyl invariants vanish. On the other hand, the affine invariants α_μ, which are not of Weyl type, completely characterize the geometry in the real analytic setting.

CHAPTER 10

Homothety Homogeneity and Local Homogeneity

The material of this chapter is based on work of García-Río, Gilkey and Nikčević [67, 68]. We also refer to recent work of Dunn and McDonald [59] on homothety homogeneity. Let $\mathcal{M}$ be a homothety homogeneity manifold which has non-trivial homotethy character, i.e., which admits a diffeomorphism ϕ so $\phi^* g = \lambda^2 g$ for $\lambda^2 \neq 1$. In Section 10.1, we show that if $\mathcal{M}$ is not VSI, then $\mathcal{M}$ is not homogeneous and present other foundational material. In Section 10.2, we give various classification results. In Section 10.3, we give a different proof of a result of Tashiro [131] that in the Riemannian setting such a manifold is necessarily incomplete. The situation is not so rigid in the Lorentzian case where pp-wave metrics support non-Killing homothety vector fields (see, for example, Alekseevski [3], Kühnel and Rademacher [94], or Steller [127] and references therein). In Section 10.4, we examine a family of Lorentzian Walker manifolds. We determine which elements of the family are 0-curvature homogeneous, which are 1-curvature homogeneous, and which are 2-curvature homogeneous; for this family, 2-curvature homogeneity will in fact imply local homogeneity. A similar analysis of homothety curvature homogeneity will be performed in Section 10.5. This will illustrate some of the more theoretical material of previous sections. In Section 10.6, we change direction slightly to prove a stability result that lets us pass from the algebraic to the geometric level by showing that k-curvature homogeneity for $k = \frac{1}{2}m(m-1)$ implies local homogeneity. In Section 10.7, we examine locally homogeneous metric G-structures in a quite general context.

10.1 INTRODUCTION

We will discuss the homothety character, homothety vector fields, the homothety short exact sequence, and k-homothety curvature homogeneity.

10.1.1 THE HOMOTHETY CHARACTER. Let $\mathcal{D}(M)$ be the group of diffeomorphisms of a pseudo-Riemannian manifold $\mathcal{M} = (M, g)$. The *homothety group* $\mathcal{H}(M)$ and the *isometry group* $\mathcal{I}(\mathcal{M})$ are defined by setting:

$$\begin{aligned} \mathcal{H}(\mathcal{M}) &:= \{\phi \in \mathcal{D}(M) : \exists\, \lambda(\phi) > 0 : \phi^* g = \lambda(\phi)^2 g\}, \\ \mathcal{I}(\mathcal{M}) &:= \{\phi \in \mathcal{H}(M) : \phi^* g = g\}. \end{aligned} \tag{10.1.a}$$

A group homomorphism from a group $\mathcal{G}$ to $\mathbb{R}^+$ is called a *multiplicative character* of $\mathcal{G}$. The scale factor $\lambda(\phi)$ of Equation (10.1.a) is called the *homothety character* since

$$\lambda(\phi_1\phi_2) = \lambda(\phi_1)\lambda(\phi_2) \quad \text{for} \quad \phi_1, \phi_2 \in \mathcal{H}(\mathcal{M}).$$

Because $\mathcal{I}(\mathcal{M}) = \ker\{\lambda\}$, $\mathcal{I}(\mathcal{M})$ is a normal subgroup of $\mathcal{H}$. We say that $\mathcal{H}$ has a *non-trivial homothety character* if λ is non-trivial or, equivalently, if $\mathcal{I}(\mathcal{M}) \neq \mathcal{H}(\mathcal{M})$ so that there exist non-trivial homotheties of $\mathcal{M}$. We say that $\mathcal{M}$ is *homothety homogeneous* (resp. *homogeneous*) if $\mathcal{H}(\mathcal{M})$ (resp. $\mathcal{I}(\mathcal{M})$) acts transitively on M. There are similar local notions where the diffeomorphism ϕ is not assumed globally defined.

10.1.2 WEYL SCALAR INVARIANTS. We recall some material from Section 9.2. Let $\mathcal{W}$ be a Weyl scalar invariant of order ℓ (i.e., $\mathcal{W}$ involves a total of ℓ derivatives of the metric tensor). Such invariants are constructed from the covariant derivatives of the curvature tensor by contracting indices in pairs. For example, the scalar curvature

$$\tau := g^{i\ell} g^{jk} R_{ijk\ell}$$

is an invariant of order 2. Recall that $\mathcal{M}$ is said to be VSI (see Section 9.5) if all the scalar Weyl invariants of $\mathcal{M}$ vanish. In the Riemannian setting, $\mathcal{M}$ is VSI if and only if $\mathcal{M}$ is flat. This is not true in the pseudo-Riemannian setting as Example 9.17 demonstrates. The following is a useful observation that will play a central role in our discussion.

Lemma 10.1 Let $\mathcal{M} = (M, g)$ be a pseudo-Riemannian manifold which is not VSI. If the homothety character of $\mathcal{M}$ is non-trivial, then $\mathcal{M}$ is not homogeneous.

Proof. Choose $\phi \in \mathcal{H}(\mathcal{M})$ so that $\phi^* g = \lambda^2 g$ for $\lambda > 0$ and $\lambda \neq 1$. Let $\mathcal{W}$ be Weyl scalar invariant of order ℓ which does not vanish identically on $\mathcal{M}$. By Lemma 9.26,

$$\phi^*(\mathcal{W}) = \lambda^{-\ell}\mathcal{W}. \tag{10.1.b}$$

Thus, $\mathcal{W}(\mathcal{M})$ is not constant and, consequently, $\mathcal{M}$ is not homogeneous. □

Lemma 10.1 can fail in the VSI setting.

Lemma 10.2 Let $(x, y, \tilde{x}, \tilde{y})$ be coordinates on $\mathbb{R}^4$. Let $\mathcal{M} = (\mathbb{R}^4, g)$ where g is the metric of signature $(2, 2)$ given by $ds^2 = e^{2y} dx \otimes dx + dx \otimes d\tilde{x} + d\tilde{x} \otimes dx + dy \otimes d\tilde{y} + d\tilde{y} \otimes dy$. Then $\mathcal{M}$ is homogeneous, non-flat, geodesically complete, and has non-trivial homothety character.

Note that $\mathcal{M}$ has the form given in Example 9.18 with $\ell = -1$ and $f(y)$ chosen suitably.

Proof. For $\lambda > 0$, let $S_\lambda(x, y, \tilde{x}, \tilde{y}) := (\lambda x, y, \lambda\tilde{x}, \lambda^2\tilde{y})$. Then $S_\lambda^* g = \lambda^2 g$ so the homothety character of $\mathcal{M}$ is non-trivial. Define

$$T_{\vec{a}}(x, y, \tilde{x}, \tilde{y}) := (e^{-a_2}x + a_1, y + a_2, e^{a_2}\tilde{x} + a_3, \tilde{y} + a_4)$$

for $\vec{a} = (a_1, a_2, a_3, a_4) \in \mathbb{R}^4$. Since the transformations $\{T_{\vec{a}}\}$ act transitively on $\mathcal{M}$ by isometries, $\mathcal{M}$ is homogeneous. The non-zero covariant derivatives are

$$\begin{aligned} g(\nabla_{\partial_x}\partial_x, \partial_y) &= -e^{2y}, & g(\nabla_{\partial_x}\partial_y, \partial_x) &= g(\nabla_{\partial_y}\partial_x, \partial_x) = e^{2y}, \\ \nabla_{\partial_x}\partial_x &= -e^{2y}\partial_{\tilde{y}}, & \nabla_{\partial_x}\partial_y &= \nabla_{\partial_y}\partial_x = e^{2y}\partial_{\tilde{x}}\,. \end{aligned}$$

This shows $\mathcal{M}$ is a generalized plane wave manifold. Consequently, by Theorem 9.46, $\mathcal{M}$ is VSI and geodesically complete. Furthermore, $\mathcal{M}$ is not flat since

$$\begin{aligned} \mathcal{R}(\partial_y, \partial_x)\partial_x &= \{\nabla_{\partial_y}\nabla_{\partial_x} - \nabla_{\partial_x}\nabla_{\partial_y}\}\partial_x = \nabla_{\partial_y}(-e^{2y}\partial_{\tilde{x}}) - \nabla_{\partial_x}(e^{2y}\partial_{\tilde{x}}) \\ &= -2e^{2y}\partial_{\tilde{x}}\,. \end{aligned}$$

□

Let $\mathcal{M}$ be a homothety homogeneous manifold which is not VSI. Let P_0 be the base point of $\mathcal{M}$. We will use P_0 to normalize our invariants; the choice is inessential. Choose a scalar Weyl invariant $\mathcal{W}$ which does not vanish identically on M. Given any point $P \in M$, choose $\phi_P \in \mathcal{H}(\mathcal{M})$ so that $\phi_P(P_0) = P$. By Equation (10.1.b), $\mathcal{W}(P_0) = \lambda(\phi_P)^{-\ell}\mathcal{W}(P)$. Thus, since $\mathcal{W}$ does not vanish identically, $\mathcal{W}$ never vanishes. Define

$$\mu_{\mathcal{W}}(P) := \left|\frac{\mathcal{W}(P_0)}{\mathcal{W}(P)}\right|^{1/\ell} \quad \text{and} \quad \mathcal{M}_c^{\mathcal{W}} := \{P \in M : \mu_{\mathcal{W}}(P) = c\}\,.$$

Theorem 10.3 *Let $\mathcal{M} = (M, g)$ be a homothety homogeneous non VSI manifold with non-trivial homothety character λ. Then the level sets $\mathcal{M}_c^{\mathcal{W}}$ are smooth submanifolds of M of codimension 1 which are independent of the particular non-vanishing Weyl scalar invariant $\mathcal{W}$ which was chosen. $\mathcal{I}$ acts transitively on these level sets so $\mathcal{M}$ has cohomogeneity 1.*

Proof. Let $\phi \in \mathcal{H}$. By Equation (10.1.b), $\phi^*\mu_{\mathcal{W}} = \lambda(\phi)\mu_{\mathcal{W}}$ and thus $\phi^* d\mu_{\mathcal{W}} = \lambda(\phi)d\mu_{\mathcal{W}}$. Since λ is non-trivial, $\mu_{\mathcal{W}}$ is non-constant. Thus, $d\mu_{\mathcal{W}}$ does not vanish identically so $d\mu_{\mathcal{W}}$ is never zero and the level sets $M_c^{\mathcal{W}}$ are smooth submanifolds of M which have codimension 1. Furthermore,

$$\phi : M_c^{\mathcal{W}} \to M_{\lambda(\phi)c}^{\mathcal{W}} \quad \text{for any} \quad \phi \in \mathcal{H}\,.$$

Since $\mathcal{H}$ acts transitively on M, $\mathcal{I} := \ker\{\lambda\}$ acts transitively on $M_c^{\mathcal{W}}$. Thus $M_1^{\mathcal{W}} = \mathcal{I} \cdot P_0$ and $M_c^{\mathcal{W}} = \phi \cdot \mathcal{I} \cdot P_0$ for any $\phi \in \mathcal{H}$ with $\lambda(\phi) = c$. This shows the level sets $M_c := M_c^{\mathcal{W}}$ are in fact independent of the choice of $\mathcal{W}$. □

Example 10.4 The induced metric on the level sets can be degenerate in the pseudo-Riemannian context. Let $\mathcal{N} = (N, g_N)$ be a homogeneous manifold which is not VSI. Let

$$M := \mathbb{R}^2 \times N \quad \text{and} \quad g_{M,\varepsilon} := e^{\varepsilon x^1}(dx^1 \otimes dx^2 + dx^2 \otimes dx^1 + g_N)\,.$$

The shift $(x^1, x^2, \xi) \to (x^1, x^2 + v, \xi)$ and the isometries of $\mathcal{N}$ act transitively on the slices $\{x^1\} \times \mathbb{R} \times N$. Furthermore, the shift $(x^1, x^2, \xi) \to (x^1 + u, x^2, \xi)$ is a homothety of $\mathcal{M}$ and rescales the metric by $e^{\varepsilon u}$. Consequently, $\mathcal{M}$ is homothety homogeneous with non-trivial homothety character. Let $\mathcal{W}$ be a non-trivial Weyl scalar invariant of $\mathcal{N}$. Setting $\epsilon = 0$ yields a product metric of $\mathcal{N}$ with a flat factor. Consequently, $\mathcal{W}(g_{M,0}) = \mathcal{W}(g_N) \neq 0$. Since $\mathcal{W}(g_{M,\varepsilon})$ is a real analytic function of ε, $\mathcal{M}_\varepsilon$ is not VSI for generic $\varepsilon \neq 0$. The level sets $\mathcal{M}_c$ are given by $x^1 = \tilde{c}$ for $\tilde{c} = \tilde{c}(c)$. These slices inherit a degenerate metric.

10.1.3 HOMOTHETY VECTOR FIELDS. Let $\mathcal{L}$ be the *Lie derivative*. Recall that a vector field X is said to be a *Killing vector field* if $\mathcal{L}_X g = 0$ or, equivalently as was shown in Lemma 7.8 of Book II, if $X_{a;b} + X_{b;a} = 0$ where ";" represents covariant differentiation with respect to the Levi–Civita connection. In Lemma 9.49, we discussed affine Killing vector fields. The equivalent result in the homothety setting is given as follows:

Lemma 10.5 Let $\mathcal{M} = (M, g)$ be a pseudo-Riemannian manifold.

1. Suppose that X is a smooth vector field on M. If any of the following equivalent conditions is satisfied, then X is said to be a *homothety vector field*:
   (a) $\mathcal{L}_X g = cg$ for some $c \in \mathbb{R}$.
   (b) The local 1-parameter flows ϕ_t^X are local homotheties.
   (c) $X_{a;b} + X_{b;a} = \lambda g_{ab}$ for some constant λ.
2. If X and Y are homothety vector fields, then $[X, Y]$ is a Killing vector field.

Proof. Since the local flows are local homotheties, $(\phi_t^X)^* g = \mu_t^2 g$. Since $\phi_t^X \phi_s^X = \phi_{t+s}^X$, we have $\mu_t^2 \mu_s^2 = \mu_{s+t}^2$ and, consequently, $\mu_t^2 = e^{ct}$ for some constant c so $(\phi_t^X)^* g = e^{ct} g$. Differentiating this relation with respect to t and setting $t = 0$ then yields $\mathcal{L}_X g = cg$. Conversely, suppose $\mathcal{L}_X g = cg$. If $X(P) \neq 0$, then we may choose local coordinates $\vec{x} = (x^1, \dots, x^m)$ so $X = \partial_{x^1}$. The condition $\mathcal{L}_X g = cg$ implies that $g_{ij}(x^1 + t, \dots, x^m) = e^{ct} g_{ij}(x^1, \dots, x^m)$ and, consequently, the shift $\phi_t^X(x^1, \dots, x^m) \to (x^1 + t, x^2, \dots, x^m)$ is a homothety; continuity then yields this result even if $X(P) = 0$ (see the argument used to prove Lemma 7.7 in Book II). Thus, Assertion 1-a and Assertion 1-b are equivalent. We use the same argument used to prove Lemma 7.8 in Book II in examining the equivalence with Assertion 1-c. Let $A_X Y := \nabla_Y X$. We compute:

$$\begin{aligned}(\mathcal{L}_X g)(Y, Z) &= (\mathcal{L}_X g)(Y \otimes Z) = X \cdot g(Y \otimes Z) - g(\mathcal{L}_X(Y \otimes Z))\\ &= X \cdot g(Y, Z) - g([X, Y], Z) - g([X, Z], Y)\\ &= X \cdot g(Y, Z) - g(\nabla_X Y, Z) + g(\nabla_Y X, Z) - g(\nabla_X Z, Y) + g(\nabla_Z Y, X)\\ &= (\nabla_X g)(Y, Z) - g(A_X Y, Z) - g(A_X Z, Y)\\ &= -\{g(A_X Y, Z) + g(A_X Z, Y)\}.\end{aligned}$$

Let $Y = \partial_{x^a}$ and $Z = \partial_{x^b}$. We then have

$$\mathcal{L}_X g(\partial_{x^a}, \partial_{x^b}) = -g(\nabla_{\partial_a} X, \partial_b) - g(\nabla_{\partial_b} X, \partial_a) = -X_{b;a} - X_{a;b}\,.$$

Thus, if $\mathcal{L}_X g = -cg$, then $-cg = -X_{b;a} - X_{a;b}$. Conversely, if $-cg = -X_{b;a} - X_{a;b}$, then $\mathcal{L}_X g = -cg$. This establishes the equivalence of Assertion 1-a and Assertion 1-c.

By Lemma 7.7 of Book II, $[\mathcal{L}_X, \mathcal{L}_Y] = \mathcal{L}_{[X,Y]}$. We prove Assertion 2 by computing:

$$\begin{aligned}\mathcal{L}_{[X,Y]} g &= \mathcal{L}_X \mathcal{L}_Y g - \mathcal{L}_Y g \mathcal{L}_X g = \mathcal{L}_X(\lambda_Y^2 g) - \mathcal{L}_Y(\lambda_X^2 g)\\ &= \lambda_X^2 \lambda_Y^2 g - \lambda_Y^2 \lambda_X^2 g = 0\,.\end{aligned}$$

□

10.1.4 THE HOMOTHETY SHORT EXACT SEQUENCE. Adopt the notation of Equation (10.1.b). Since $\mathcal{I} = \ker\{\lambda\}$, $\mathcal{I}$ is a normal subgroup of $\mathcal{H}$ and λ is an isomorphism from $\mathcal{H}/\mathcal{I}$ to $\mathbb{R}^+$. Let $\mathfrak{H}$ (resp. $\mathfrak{I}$) be the Lie algebra of $\mathcal{H}$ (resp. $\mathcal{I}$); elements of $\mathfrak{H}$ are homothety vector fields and elements of $\mathfrak{I}$ are Killing vector fields. The short exact sequence

$$1 \to \ker\{\lambda\} \to \mathcal{H} \xrightarrow{\lambda} \mathbb{R}^+ \to 1$$

which is defined by the homothety character is called the *homothety short exact sequence*. This sequence is said to *split* if $\mathcal{H}$ is isomorphic as a Lie group to $\mathbb{R}^+ \times \mathcal{I}$ and if under this isomorphism, λ is projection on the first factor.

Lemma 10.6 The homothety short exact sequence is split if and only if there exists a *splitting homothety vector field* $0 \neq X_{\mathfrak{H}} \in \mathfrak{H} - \mathfrak{I}$ so that $[X_{\mathfrak{H}}, Y] = 0$ for all $Y \in \mathfrak{H}$. Such a $X_{\mathfrak{H}}$ generates a 1-parameter subgroup ϕ_t such that $\lambda(\phi_t) = t$ for $t \in \mathbb{R}^+$ which gives rise to the splitting.

Proof. Suppose first that $\mathcal{H} = \mathbb{R}^+ \times \mathcal{I}$. Let $X = x\partial_x$ be the invariant vector field on $\mathbb{R}^+$. Let $\phi_x : P \to x \cdot P$ be the associated transformation. Then by assumption, $\phi_x^* g = x^2 g$. Thus, the associated vector field on M, which we continue to denote by $x\partial_x$, is a non-trivial homothety vector field which commutes with all the homotheties of $\mathcal{M}$ and, consequently, $[X, Y] = 0$ for all $Y \in \mathfrak{h}$. Conversely, if $X \in \mathfrak{h}$ has $[X, Y] = 0$ for all $Y \in \mathfrak{h}$, then the associated 1-parameter flow $\phi_t : \mathbb{R} \to \mathcal{H}$ is central in $\mathcal{H}$. We set $x = e^t$ to replace $\mathbb{R}$ by $\mathbb{R}^+$ and ∂_t by $x\partial_x$ to replace x by e^x and ∂_x by $x\partial_x$ to complete the proof. □

In the Riemannian setting, if $\mathcal{I}$ is compact and connected, then we can average over the group $\mathcal{I}$ to construct a splitting homothety vector field. The following example shows that the homothety sequence does not always split.

Example 10.7 We generalize the construction of Theorem 9.20. For $n \geq 2$, let $\mathcal{T}_n$ be the group of upper triangular $n \times n$ real matrices with positive entries on the diagonal:

$$\mathcal{T}_n := \left\{ \phi \in \mathrm{GL}(n, \mathbb{R}) : \phi = \begin{pmatrix} \phi_{11} & \phi_{12} & \dots & \phi_{1n} \\ 0 & \phi_{22} & \dots & \phi_{2n} \\ \dots & \dots & \dots & \dots \\ 0 & 0 & \dots & \phi_{nn} \end{pmatrix} \quad \text{for} \quad \phi_{ii} > 0 \right\}.$$

If $0 \neq \vec{a} = (a_1, \dots, a_n) \in \mathbb{R}^n$, let $\lambda_{\vec{a}}(\phi) = \phi_{11}^{a_1} \cdot \dots \cdot \phi_{nn}^{a_n}$ be a multiplicative character of $\mathcal{T}_n$. The center of $\mathcal{T}_n$ is 1-dimensional and consists of the constant multiples of the identity $t \cdot \mathrm{Id}$. Thus, the associated sequence $1 \to \ker\{\lambda_{\vec{a}}\} \to \mathcal{T}_n \to \mathbb{R}^+ \to 1$ is split if and only $\lambda(t\,\mathrm{Id}) \neq 1$ or, equivalently, if $a_1 + \dots + a_n \neq 0$. Thus, there are non-split characters of $\mathcal{T}_n$; we could, for example, take $\lambda(\phi) = \phi_{11}\phi_{22}{}^{-1}$. Take a left-invariant pseudo-Riemannian metric g_0 on $\mathcal{T}_n$. Then $\phi^* g_0 = g_0$ for any $\phi \in \mathcal{T}_n$. Let $g(\phi) := \lambda(\phi)^2 g_0(\phi)$. Since $\phi^*(\lambda) = \lambda(\phi) \cdot \lambda$, we show that λ is the homothety character of the left-action of $\mathcal{T}_n$ on itself by computing:

$$\phi^* g = \phi^*(\lambda^2 g_0) = \phi^*(\lambda^2)\phi^* g_0 = \lambda^2(\phi)\lambda^2 g_0 = \lambda^2(\phi) g\,.$$

10.1.5 K-HOMOTHETY CURVATURE HOMOGENEITY.

By Lemma 9.21 and Remark 9.22, a pseudo-Riemannian manifold (M, g) is *k-homothety curvature homogeneous* if either of the following three equivalent conditions hold.

1. Given any two points $P, Q \in M$, there is a linear homothety $\phi = \phi_{P,Q}$ from $(T_P M, g_P)$ to $(T_Q M, g_Q)$ so that if $0 \leq \ell \leq k$, then $\phi^*(\nabla^\ell \mathcal{R}_Q) = \nabla^\ell \mathcal{R}_P$.
2. Given any two points $P, Q \in M$, there is a linear isometry $\phi = \phi_{P,Q}$ from $T_P M$ to $T_Q M$ and $0 \neq \lambda = \lambda_{P,Q} \in \mathbb{R}$ so that if $0 \leq \ell \leq k$, then $\phi^*(\nabla^\ell R_Q) = \lambda^{-\ell-2}\nabla^\ell R_P$.
3. There exists a k-homothety curvature model for $\mathcal{M}$.

In Example 9.17 and Example 9.18, we exhibited examples which were k-curvature homogeneous but not affine $(k+1)$-curvature homogeneous for any k. Thus, they are also k-homothety curvature homogeneous but not $(k+1)$-homothety curvature homogeneous. They were all of higher signature. In the following result, we apply the construction of Theorem 9.20 to the Riemannian setting; the compactness of the orthogonal group is an essential ingredient in the proof.

Lemma 10.8 Fix $k \geq 0$. Let $\mathcal{N} = (N, g_N)$ be a k-curvature homogeneous Riemannian manifold which is not $(k+1)$-curvature homogeneous. Let

$$M := \mathbb{R} \times N, \quad g_t := e^{tx}(dx^2 + g_N) \quad \text{and} \quad \mathcal{M}_t := (M, g_t).$$

1. g_t is k-curvature homogeneous for any t.
2. There exists $\varepsilon > 0$ so that if $0 < t < \varepsilon$, then $\mathcal{M}_t$ is neither $(k+1)$-homothety curvature homogeneous nor 0-curvature homogeneous.

Proof. Fix t. Let $x \in \mathbb{R}$ and let $P \in N$. Take geodesic coordinates on N centered at P. Then the derivatives of the metric for g_N up to order $k+2$ are determined by the curvature tensor of N. Thus, the derivatives of the metric for g_t up to order $k+2$ are determined as well by (x, t) and the curvature of N. Consequently, the curvature tensor and all its covariant derivatives up to order $k+2$ at $(x, P) \in M$ are given canonically and universally by (x, t) and by the covariant derivatives of the curvature tensor up to order $k+2$ of $\mathcal{N}$ at P. Let $P, Q \in N$. Since $\mathcal{N}$ is assumed k-curvature homogeneous, there is an isometry $\Phi_{P,Q}$ from $T_P N$ to $T_Q N$ so that $\Phi_{P,Q}^* \nabla^i R_Q^{\mathcal{N}} = \nabla^i R_P^{\mathcal{N}}$ for $0 \le i \le k$. Extend $\Phi_{P,Q}$ to a linear isometry from $T_{x,P} M$ to $T_{x,Q} M$ so $\Phi_{P,Q}(\partial_x) = \partial_x$. We then obtain by naturality that $\Phi_{P,Q}^* \nabla^i R_{(x,P)}^{\mathcal{M}_t} = \nabla^i R_{(x,Q)}^{\mathcal{M}_t}$ for $0 \le i \le k$. On the other hand, the maps $\phi_c(x, P) = (x + c, P)$ are globally defined homotheties which take the slice $\{x\} \times N$ to the slice $\{x + c\} \times N$. It now follows that $\mathcal{M}_t$ is k-homothety curvature homogeneous for any t.

Suppose there exists a sequence $t_n \to 0$ so that $\mathcal{M}_{t_n}$ is $(k+1)$-homothety curvature homogeneous. Since $\mathcal{N}$ is not $(k+1)$-curvature homogeneous, we may find points P and Q in N so there is no linear isometry Φ from $T_P N$ to $T_Q N$ satisfying $\Phi^* \nabla^i R_Q^{\mathcal{N}} = \nabla^i R_P^{\mathcal{N}}$ for $0 \le i \le k+1$. Note that the metrics on $T_{(0,P)} M$ and $T_{(0,Q)} M$ are independent of t. By assumption, $\mathcal{M}_{t_n}$ is $(k+1)$-homothety curvature homogeneous. Thus, we may choose isometries Φ_n from $T_{(0,P)} M$ to $T_{(0,Q)} M$ so

$$\Phi_n^* \nabla^i R_{(0,Q)}^{\mathcal{M}_{t_n}} = \lambda_n^{-2-i} \nabla^i R_{(0,P)}^{\mathcal{M}_{t_n}} \quad \text{for} \quad 0 \le i \le k+1. \tag{10.1.c}$$

Since $g_t(0, P)$ (resp. $g_t(0, Q)$) is independent of t, all the maps Φ_n belong to a fixed compact set. Thus by passing to a subsequence, we may assume that the sequence Φ_n converges to an isometry Φ_0 from $T_{(0,P)} M$ to $T_{(0,Q)} M$. Let $\|R\|^2$ be the norm of the curvature tensor.

1. By Equation (10.1.c), $\|R_{(0,Q)}^{\mathcal{M}_{t_n}}\|^2 = \lambda_n^{-4} \|R_{(0,P)}^{\mathcal{M}_{t_n}}\|^2$.
2. Since $\mathcal{N}$ is 0-curvature homogeneous and not flat, $\|R_Q^{\mathcal{N}}\|^2 = \|R_P^{\mathcal{N}}\|^2 \neq 0$.
3. If $t = 0$, the metric is flat in ∂_x so $\|R_{(0,Q)}^{\mathcal{M}_0}\|^2 = \|R_Q^{\mathcal{N}}\|^2 = \|R_P^{\mathcal{N}}\|^2 = \|R_{(0,P)}^{\mathcal{M}_0}\|^2$.
4. $\|R_{(0,Q)}^{\mathcal{M}_{t_n}}\|^2$ (resp. $\|R_{(0,P)}^{\mathcal{M}_{t_n}}\|^2$) converges to $\|R_Q^{\mathcal{N}}\|^2$ (resp. $\|R_P^{\mathcal{N}}\|^2$) as $t_n \to 0$.

This implies that λ_n^{-4} converges to 1. Consequently,

$$\Phi_0^* \nabla^i R_{(0,Q)}^{\mathcal{M}_0} = \nabla^i R_{(0,P)}^{\mathcal{M}_0} \quad \text{for} \quad 0 \le i \le k+1.$$

Let $\mathcal{K}_{k+1}(\mathcal{N}, P)$ (resp. $\mathcal{K}_{k+1}(\mathcal{M}, (0, P))$) be the vector space of all tangent vectors $\xi \in T_P N$ (resp. $\xi \in T_{(0,P)} M$) so that if ξ is inserted into any argument of $\nabla^i R_P^{\mathcal{N}}$ (resp. $\nabla^i R_{(0,P)}^{\mathcal{M}_0}$), then one gets 0 for $0 \leq i \leq k+1$ for any choice of the remaining vectors. We note that

$$\begin{aligned}
&\mathcal{K}_{k+1}(\mathcal{M}_0,(0,P)) = \partial_x \cdot \mathbb{R} \oplus \mathcal{K}_{k+1}(\mathcal{N},P), \qquad \text{and}\\
&\Phi_0^*\{\mathcal{K}_{k+1}(\mathcal{M}_0,(0,Q))\} = \{\mathcal{K}_{k+1}(\mathcal{M}_0,(0,P))\}\,.
\end{aligned}$$

Thus, Φ_0 induces an isometry $\Phi_0 : \mathcal{K}_{k+1}(\mathcal{M}_0(0,P))^\perp \to \mathcal{K}_{k+1}(\mathcal{M}_0(0,Q))^\perp$. We have

$$\begin{aligned}
\dim(\mathcal{K}_{k+1}(\mathcal{N},P)) &= \dim(\mathcal{K}_{k+1}(\mathcal{M}_0,(0,P))) - 1\\
&= \dim(\mathcal{K}_{k+1}(\mathcal{M}_0,(0,Q))) - 1 = \dim(\mathcal{K}_{k+1}(\mathcal{N},Q))\,.
\end{aligned}$$

Let Ψ_0 be the restriction of Φ_0 to $\mathcal{K}_{k+1}(\mathcal{M}_0,(0,P))^\perp$. Extend Ψ_0 to an isometry from $\mathcal{K}_{k+1}(\mathcal{N},P)$ to $\mathcal{K}_{k+1}(\mathcal{N},Q)$ arbitrarily. Then $\Psi_0^*\nabla^i R_Q^{\mathcal{N}} = \nabla^i R_P^{\mathcal{N}}$ for $0 \leq i \leq k+1$. This contradicts the choice of P and Q and shows $\mathcal{M}_t$ is not $(k+1)$-homothety curvature homogeneous for small t. Finally, we use Theorem 9.20 to see that the scalar curvature of $\mathcal{M}_t$ is non-constant for generic t and, therefore, $\mathcal{M}_t$ is not 0-curvature homogeneous. □

10.2 CLASSIFICATION RESULTS

In this section, we give several classification results. In Section 10.2.1, we give a general classification result of homothety homogeneous manifolds with non-trivial homothety character which are not VSI. In Section 10.2.2, we examine classification results for the special case that the homothety exact sequence is split and characterize abstractly the examples appearing in Theorem 9.20.

10.2.1 A GENERAL CLASSIFICATION RESULT.

Definition 10.9 Let $\mathcal{M}_0 := (M, g_0)$ be a homogeneous pseudo-Riemannian manifold. Let $\mathcal{J}$ be a subgroup of the group of isometries $\mathcal{I}(\mathcal{M}_0)$ which acts transitively on M. Let $\mathcal{J}_0$ be the isotropy subgroup of a point $P \in M$ so that $M = \mathcal{J}/\mathcal{J}_0$. Assume there exists a non-trivial multiplicative character λ of $\mathcal{J}$ so that $\mathcal{J}_0 \subset \ker\{\lambda\}$. If $P \in M$, choose $\psi_P \in \mathcal{J}$ so $\psi_P(P_0) = P$ and define $\lambda(P) = \lambda(\psi_P)$. If $\tilde\psi_P \in \mathcal{J}$ satisfies $\tilde\psi_P(P_0) = P$, then $\psi_P^{-1}\tilde\psi_P P_0 = P_0$. Consequently, we have $\psi_P^{-1}\tilde\psi_P \mathcal{J}_0 = \text{Id}$. This implies that $\lambda(\psi_P^{-1}\tilde\psi_P) = 1$ and shows that $\lambda(\psi) = \lambda(\tilde\psi)$ and, therefore, $\lambda(P)$ is a smooth well-defined map from M to $\mathbb{R}^+$. Let

$$\mathcal{M} = \mathcal{M}(\mathcal{M}_0, \mathcal{J}, \lambda) := (M, \lambda^2 g_0).$$

Theorem 10.10 *Adopt the notation established above.*

1. $\mathcal{M}(\mathcal{M}_0, \mathcal{J}, \lambda)$ *is homothety homogeneous with non-trivial homothety character* λ.
2. *If* $\tilde{\mathcal{M}} = (\tilde M, \tilde g)$ *is a homothety homogeneous manifold with non-trivial homothety character which is not VSI, then* $\tilde{\mathcal{M}} = \mathcal{M}(\mathcal{M}_0, \mathcal{J}, \lambda)$ *for some* $(\mathcal{M}_0, \mathcal{J}, \lambda)$.

Proof. Let $\mathcal{M} = \mathcal{M}(\mathcal{M}_0, \mathcal{J}, \lambda)$. Let $\phi \in \mathcal{J}$. As $\mathcal{J}$ acts by isometries on $\mathcal{M}_0$, $\phi^* g_0 = g_0$. Consequently, $\phi^*(\lambda^2 g_0) = \phi^*(\lambda^2)\phi^*(g_0) = \phi^*(\lambda^2)g_0$. Given $P \in M$, chose ψ so $\psi(P_0) = P$. We then have $\lambda(P) = \lambda(\psi)$. Thus, $\lambda(\phi P) = \lambda(\phi\psi) = \lambda(\phi)\lambda(\psi) = \lambda(\phi)\lambda(P)$. Consequently, $\phi^*\lambda = \phi(\lambda)\lambda$ and $\phi^*(\lambda^2 g) = \lambda(\phi)^2(\lambda^2 g)$. Assertion 1 now follows.

Conversely, suppose that (M, g) is homothety homogeneous with non-trivial homothety character λ and that (M, g) is not VSI. We can use Theorem 10.3 to define the level sets M_c. We set $g_0(P) := \mu_{\mathcal{W}}(P)^{-2} g(P)$ to define a conformally equivalent manifold on which $\mathcal{J} := \mathcal{I}(\mathcal{M})$ acts by isometries. The corresponding structure of Definition 10.9 is then given by $((M, g_0), \mathcal{J}, \lambda)$. □

10.2.2 SPLIT HOMOTHETY EXACT SEQUENCE.

Definition 10.11 Let $\mathcal{I}$ be a connected Lie group which acts transitively and effectively by isometries on a connected pseudo-Riemannian manifold $\mathcal{N} = (N, g_N)$. Assume given a smooth 1-form θ on N which is invariant under the action of $\mathcal{I}$. Let $M := \mathbb{R} \times N$ and let

$$g^{\pm}_{M,t} := e^{tx}(\pm dx^2 + dx \otimes \theta + \theta \otimes dx + ds^2_N) \quad \text{and} \quad g^0_{M,t} := e^{tx}(dx \otimes \theta + \theta \otimes dx + ds^2_N)\,.$$

The normalization is chosen for $g^+_{M,t}$ (resp. $g^-_{M,t}, g^0_{M,t}$) so that the homothety vector field ∂_x is a unit spacelike (resp. timelike or null) vector field; the parameter t then reflects the fact that the homothety character for the translation $(x, P) \to (x + c, P)$ is e^{tc}.

Lemma 10.12 Adopt the notation of Definition 10.11.

1. The metric $g^{\pm}_{M,t}$ is non-degenerate if and only if $g_N(\theta, \theta) \neq \pm 1$.
2. The metric $g^0_{M,t}$ is non-degenerate if and only if $g_N(\theta, \theta) \neq 0$.

Proof. We first prove Assertion 1. We assume ∂_x is spacelike as the timelike case is analogous. If $\theta = 0$, then $g^+_{M,t}$ is non-singular so we suppose $\theta \neq 0$. Fix a point P of N; the particular point P being irrelevant as $\mathcal{N}$ is homogeneous.

Case 1. Let $g = g^+_{M,t}$. Suppose that θ is not a null covector. Choose an orthonormal basis $\{Y_2, \dots, Y_m\}$ for $T_P N$ so that $\theta(Y_i) = 0$ for $i \geq 3$, so that $\theta(Y_2) = c$, and so that

$$g(Y_i, Y_j) = \epsilon_i \delta_{ij} \quad \text{where} \quad \epsilon_i = \pm 1 \quad \text{for} \quad 2 \leq i, j \leq m\,.$$

Then $g(\theta, \theta) = c^2 \epsilon_2$. Let $Y_1 = \partial_x$. We then have:

$$\det(g_{ij}) = e^{mtx} \det \begin{pmatrix} 1 & c \\ c & \epsilon_2 \end{pmatrix} \cdot \prod_{i \geq 3} \epsilon_i\,. \tag{10.2.a}$$

If $\epsilon_2 = -1$, then $\det(g_{ij}) \neq 0$ and $g_N(\theta, \theta) < 0$. If $\epsilon_2 = +1$, then g is non-degenerate if and only if $c^2 \neq 1$ or, equivalently, if and only if $g_N(\theta, \theta) \neq 1$.

Case 2. Let $g = g^{+}_{M,t}$. Suppose $0 \neq \theta$ is a null covector. We can choose an orthonormal basis $\{Y_2, \dots, Y_m\}$ for $T_P N$ so Y_2 is spacelike, Y_3 is timelike, $\theta(Y_2) = \theta(Y_3) = c$, and $\theta(Y_i) = 0$ for $i \geq 4$. We then have

$$\det(g_{ij}) = e^{mtx} \det \begin{pmatrix} 1 & c & c \\ c & 1 & 0 \\ c & 0 & -1 \end{pmatrix} \cdot \prod_{i \geq 4} \epsilon_i = -e^{mtx} \cdot \prod_{i \geq 4} \epsilon_i \neq 0 . \tag{10.2.b}$$

We now turn to the proof of Assertion 2 in which ∂_x is a null vector. Let $g = g^{0}_{M,t}$. If θ is not a null-covector, then the analysis of Case 1 pertains and the same calculation as that used to derive Equation (10.2.a) shows

$$\det(g_{ij}) = e^{mtx} \det \begin{pmatrix} 0 & c \\ c & \epsilon_2 \end{pmatrix} \cdot \prod_{i \geq 3} \epsilon_i = e^{mtx}(-c^2) \prod_{i \geq 3} \epsilon_i \neq 0 .$$

On the other hand, if θ is a null covector, then the analysis of Case 2 pertains and the analogue of Equation (10.2.b) shows

$$\det(g_{ij}) = e^{mtx} \det \begin{pmatrix} 0 & c & c \\ c & 1 & 0 \\ c & 0 & -1 \end{pmatrix} \cdot \textstyle\prod_{i \geq 4} \epsilon_i = 0 . \qquad \square$$

Lemma 10.13 Let $\mathcal{M}$ be as in Definition 10.11. Assume the non-degeneracy condition of Lemma 10.12 holds. Then $\mathcal{M}$ is homothety homogeneous with non-trivial homothety character.

1. $\mathcal{M}$ is isomorphic to a manifold with $\theta = 0$ if and only if θ is dual to a Killing vector field, i.e., $\theta_{a;b} + \theta_{b;a} = 0$ for all (a, b).
2. If $\theta = 0$ and if $\tau_{\mathcal{N}} - \frac{(m-1)(m-2)}{4} t^2 \neq 0$, then $\mathcal{M}$ is not 0-curvature homogeneous.

Proof. To prove Assertion 1, we note that $\mathcal{N}$ is homogeneous so the isometries of $\mathcal{N}$ act transitively on the slices $\{x\} \times N$. The translations $(x, \xi) \to (x + c, \xi)$ are homotheties and, consequently, $\mathcal{M}$ is homothety homogeneous. Since $t \neq 0$, the homothety constant of such a translation is non-trivial for $c \neq 0$.

To prove Assertion 2, we suppose first that $\mathcal{M}$ is isomorphic to a manifold $\tilde{\mathcal{M}}$ where $\theta = 0$. The slices $\{\tilde{x}\} \times N$ or $\{x\} \times N$ are the image of $\mathcal{I}$. Let $\tilde{X}$ be the homothety vector field defined by $\partial_{\tilde{x}}$; $\tilde{X}$ is central in $\mathfrak{h}$ and is perpendicular to the slices. Thus, we can write $X = c\tilde{X} + \xi$ where ξ is a Killing vector field. We then have θ is dual to a multiple of ξ with respect to the metric g_N on the slice $\{0\} \times N$ and, consequently, satisfies the equation $\theta_{a;b} + \theta_{b;a} = 0$.

On the other hand, suppose that $\theta_{a;b} + \theta_{b;a} = 0$. The associated dual vector field ξ is then a Killing vector field which is invariant under the action of $\mathcal{I}$. Since $\mathcal{I}$ acts transitively on N, we can integrate ξ to find a smooth 1-parameter flow $\{\phi_\epsilon\}_{\epsilon \in \mathbb{R}}$ of isometries which commutes with

$\mathcal{I}$. Let $\tilde{\mathcal{I}}$ be the (possibly larger) group of isometries generated by $\mathcal{I}$ and $\{\phi_\epsilon\}_{\epsilon\in\mathbb{R}}$. Denote the extended structure by $\tilde{\mathfrak{Q}} := (N, g_N, \tilde{\mathcal{I}}, 0)$. Let

$$\varrho := (1 - \|\xi\|^2)^{-1/2} \quad \text{and} \quad s := \varrho t\,.$$

Set $\tilde{\mathcal{M}}_s = \mathcal{M}_{s,\tilde{\mathfrak{Q}}}$. Define a diffeomorphism Ψ of $M = \mathbb{R} \times N$ by setting

$$\Psi_\varrho(x, y) := (x, \phi_{\varrho x} y)\,.$$

We will show that Ψ_ϱ is an isometry from $\mathcal{M}$ to $\tilde{\mathcal{M}}_s$. Fix $(x, y) \in M$. Let $Y \in T_y N$. We have that $\Psi_*\partial_x = \partial_x + \varrho\phi_{\varrho x,*}\xi$, and $\Psi_* Y = \phi_{\varrho x,*} Y$. Since $\phi_{\varrho x,*}$ is an isometry, we have:

$$\begin{aligned}
&g_{\tilde{\mathcal{M}}_s}\big(\Psi_*\partial_x, \Psi_*\partial_x\big) = e^{sx}\{1 + \varrho^2\|\xi\|^2\},\\
&g_{\tilde{\mathcal{M}}_s}\big(\Psi_*\partial_x, \Psi_* Y\big) = e^{sx}\varrho\,\theta(Y),\\
&g_{\tilde{\mathcal{M}}_s}(\Psi_* Y_1, \Psi_* Y_2) = e^{sx} g_N(Y_1, Y_2),\\
&\Psi^* g_{s,\tilde{\mathfrak{Q}}} = e^{sx}((1 + \varrho^2\|\xi\|^2)dx^2 + \varrho dx \otimes \theta + \varrho\theta \otimes dx + g_N)\,.
\end{aligned}$$

Let Υ be the diffeomorphism which changes variables as follows:

$$\tilde{x} = (1 + \rho^2\|\xi\|^2)^{1/2} x \quad \text{and} \quad \tilde{t} = (1 + \rho^2\|\xi\|^2)^{-1/2} s\,.$$

Then $\Upsilon^* g_{\tilde{\mathcal{M}}_s} = e^{\tilde{t}\tilde{x}}(d\tilde{x}^2 + \varrho(1 + \rho^2\|\xi\|^2)^{-1/2} d\tilde{x} \otimes \theta + \theta \otimes d\tilde{x} + g_N)$. We use the defining relation for ϱ to see

$$\begin{aligned}
\varrho(1 + \varrho^2\|\xi\|^2)^{-1/2} &= \left\{\frac{1}{1 - \|\xi\|^2}\right\}^{1/2} \left\{1 + \frac{\|\xi\|^2}{1 - \|\xi\|^2}\right\}^{-1/2} = 1,\\
\tilde{t} = (1 + \rho^2\|\xi\|^2)^{-1/2} s &= \left(1 + \frac{\|\xi\|^2}{1 - \|\xi\|^2}\right)^{-1/2} s = (1 - \|\xi\|^2)^{1/2} s = t\,.
\end{aligned}$$

We use the variable $\tilde{x}$ instead of x to see that Ψ provides an isometry between the two structures. If the conditions of Assertion 2 hold, then $\tau_{\mathcal{M}}$ is non-constant and, consequently, $\mathcal{M}$ is not 0-curvature homogeneous. □

We now establish a classification result. Let $\mathcal{M}_c$ be the slices of Theorem 10.3. Adopt the notation of Definition 10.11.

Theorem 10.14 *Let $\mathcal{M}$ be a homothety homogeneous manifold with non-trivial homothety character λ which is not VSI. Assume the induced metrics on the slices $\mathcal{M}_c$ are non-degenerate. Then the homothety sequence splits if and only if $\mathcal{M}$ has the form given in Definition 10.11.*

Proof. Let $\mathcal{M}$ be a homothety homogeneous manifold with non-trivial homothety character which is not VSI. Suppose the homothety character λ splits and that induced metric on the slices $\mathcal{M}_c$ is non-degenerate. Let $X_{\mathfrak{H}}$ be the splitting homothety vector field. Suppose first $X_{\mathfrak{H}}$ is spacelike. Normalize $X_{\mathfrak{H}}$ so that $g_M(X_{\mathfrak{H}}, X_{\mathfrak{H}}) = 1$ on M_1. We now write the flow additively to construct a diffeomorphism of M with $\mathbb{R} \times M_1$. Let $\theta(Y) := g_M(X_{\mathfrak{H}}, Y)$ for Y tangent to M_1. Since $X_{\mathfrak{H}}$ is invariant under the action of $\mathcal{I} = \ker\{\lambda\}$, θ is an invariant 1-form on M_1 and the metric for any point of M_1 takes the form $g_M = e^{tx}\{dx^2 + dx \otimes \theta + \theta \otimes dx + g_{M_1}\}$ as desired. If $X_{\mathfrak{H}}$ is timelike, we normalize $X_{\mathfrak{H}}$ so that $g_M(X_{\mathfrak{H}}, X_{\mathfrak{H}}) = -1$ but otherwise the analysis is similar. Finally, we suppose $X_{\mathfrak{H}}$ is null. We normalize $X_{\mathfrak{H}}$ so that $t = 1$. The rest of the analysis is the same. □

Remark 10.15 Let $N := M_1$ be the slice. It can happen, of course, that the induced inner product g_N on N is degenerate. We have $\dim(\ker\{g_N\}) = 1$ and at each point, we may choose a spanning vector field Y. Let $\theta(\cdot) := g(X_{\mathfrak{H}}, \cdot)$; since $g_1(Y, \cdot) = 0$ we have $\theta(Y) \neq 0$. Let

$$V := \ker\{\theta\} \cap TM_1 .$$

Since $g_N|_V$ is non-singular, we can express $TN = \ker\{g_1\} \oplus V$. There are three cases depending upon whether ∂_x is spacelike, timelike, or null:

1. $g_M = e^{tx}(dx^2 + dx \otimes \theta + \theta \otimes dx + g_1)$.
2. $g_M = e^{tx}(-dx^2 + dx \otimes \theta + \theta \otimes dx + g_1)$.
3. $g_M = e^{x}(dx \otimes \theta + \theta \otimes dx + g_1)$.

This gives rise to an abstract characterization of this case as well where $\mathcal{N} = (N, \theta, g_1)$ is as above and where (θ, g_1) are invariant under a transitive group action. We will not pursue this further.

10.3 COMPLETENESS

Throughout this section, let $\mathcal{M}$ be a non-flat Riemannian manifold which is homothety homogeneous with non-trivial homothety character λ. Since $\mathcal{M}$ is not flat, $\mathcal{M}$ is not VSI so the results of Section 10.1 and Section 10.2 apply. Let $\mathcal{M}_c$ be the slices of Theorem 10.3. If $P \in M$ and if $r > 0$, let $B_r(P)$ be the geodesic ball of radius r about P. Choose $\epsilon = \epsilon(P) > 0$ so that the radial geodesics in $B_\epsilon(P)$ minimize distance. Choose $\delta = \delta(P) > 0$ so that if $|c - d| < \delta(P)$, then $M_d \cap B_\epsilon(P)$ is not empty. We begin our study with the following result.

Lemma 10.16 Adopt the notation established above.

1. Let $P \in M_c$. If $|c - d| < \delta(P)$, then there is a unique point $Q \in M_d$ which is the closest point to P in M_d; $d(P, Q) = d(M_c, M_d)$. If σ is the shortest unit speed geodesic from P to Q, then σ is perpendicular to M_c at P and to M_d at Q.

2. Let $\sigma : [0, T] \to M$ be a unit speed geodesic which is perpendicular to $M_{\mu_{\mathcal{W}}(\sigma(0))}$ at $\sigma(0)$. Then σ is perpendicular to $M_{\mu_{\mathcal{W}}(\sigma(t))}$ for any t in the interval $[0, T]$. In addition, $\sigma[t_0, t_1]$ is a curve which minimizes the distance from $M_{\mu_{\mathcal{W}}(\sigma(t_0))}$ to $M_{\mu_{\mathcal{W}}(\sigma(t_1))}$ for any $0 \le t_0 < t_1 \le T$.

Proof. Let $|c - d| < \delta(P)$. Choose $Q_1 \in M_d \cap B_{\epsilon(P)}(P)$ to be a point on M_d which is a closest point to P. There might, a priori of course, be several such points. Let σ_1 be the unit speed geodesic from P to Q_1 minimizing the distance so $\sigma_1(0) = P$ and $\sigma_1(t_1) = Q_1$. If $\dot{\sigma}(t_1)$ is not perpendicular to $T_{Q_1}M_d$, then we could "cut off the leg" to construct a point Q_2 of M_d with $d(P, Q_2) < d(P, Q_1)$. As this contradicts the choice of Q_1, we must have $\dot{\sigma}(t_1) \perp T_{Q_1}M_d$. Next, suppose that $\dot{\sigma}(0)$ is not perpendicular to $T_P M_c$. Then we could "cut off the leg" to construct a point $P_1 \in M_c$ so $d(P_1, Q_1) < d(P, Q_1)$. Note that $\mathcal{I}$ acts transitively on M_c for any c. Choose $\phi \in \mathcal{I}$ so $\phi P_1 = P$. One then has that

$$d(P, \phi Q_1) = d(\phi P_1, \phi Q_1) - d(P_1, Q_1) < d(P, Q_1)$$

which again contradicts the choice of Q_1. This shows that the closest point is unique. Given any other point $P_2 \in M_c$, we construct Q_2 similarly. Choose an isometry $\phi \in \mathcal{I}$ with $\phi P_2 = P$. Then we must have that $\phi Q_2 = Q$ and thus $d(P, Q) = d(P_2, Q_2)$. This proves Assertion 1; Assertion 2 is an immediate consequence of Assertion 1. □

Theorem 10.17 *Let $\mathcal{M} = (M, g)$ be a connected Riemannian manifold which is homothety homogeneous with non-trivial homothety character λ and which is not flat. There exists a constant $\kappa = \kappa(M) > 0$ so $\operatorname{dist}(M_c, M_d) = \kappa|c - d|$. This implies that $\mathcal{M}$ is incomplete.*

Remark 10.18 Tashiro [131] showed that a complete Riemannian manifold which admits a non-homothety vector field must be flat. Consequently, a non-flat complete homothety homogeneous manifold is necessarily homogeneous in the Riemannian setting. The situation is not so rigid in the Lorentzian case where pp-wave metrics support non-Killing homothety vector fields (see, for example, Alekseevski [3], Kühnel and Rademacher [94], or Steller [127] and the references therein). The example of Lemma 10.2 is a VSI manifold of signature $(2, 2)$. Since this manifold is a generalized plane wave manifold, it is geodesically complete. The homothety takes the form $(x, y, \tilde{x}, \tilde{y}) \to (x, y + c, \tilde{x}, \tilde{y})$ and the homothety vector field ∂_y is globally defined.

Proof. Suppose $1 < s_1 < s_2$. Choose homotheties ϕ_i so $\lambda(\phi_i) = s_i$. We then have

$$\phi_1 M_1 = M_{s_1}, \quad \phi_2 M_1 = M_{s_2}, \quad (\phi_1\phi_2)M_1 = \phi_1 M_{s_2} = \phi_2 M_{s_1} = M_{s_1 s_2}.$$

Therefore,

$$\begin{aligned} d(M_1, M_{s_1 s_2}) &= d(M_1, M_{s_1}) + d(M_{s_1}, M_{s_1 s_2}) = d(M_1, M_{s_1}) + d(\phi_1 M_1, \phi_1 M_{s_2}) \\ &= d(M_1, M_{s_1}) + \lambda(\phi_1) d(M_1, M_{s_2}) = d(M_1, M_{s_1}) + s_1 d(M_1, M_{s_2}). \end{aligned}$$

Similarly, we have that $d(M_1, M_{s_1 s_2}) = d(M_1, M_{s_2}) + s_2 d(M_1, M_{s_1})$. Thus,

$$d(M_1, M_{s_1}) + s_1 d(M_1, M_{s_2}) = d(M_1, M_{s_2}) + s_2 d(M_1, M_{s_1}).$$

Consequently, $d(M_1, M_{s_1})(s_2 - 1) = d(M_1, M_{s_2})(s_1 - 1)$ so

$$\kappa := \frac{d(M_1, M_{s_1})}{s_1 - 1} = \frac{d(M_1, M_{s_2})}{s_2 - 1}$$

is independent of the choice of s_1 and s_2 for $1 < s_1 < s_2$. Let $s < t$. Choose a homothety ϕ so $\lambda(\phi) = s$. Then $\phi(M_{\frac{t}{s}}) = M_t$. Since $1 < \frac{t}{s}$,

$$d(M_s, M_t) = d(\phi M_1, \phi M_{\frac{t}{s}}) = s d(M_1, M_{\frac{t}{s}}) = s\kappa(\tfrac{t}{s} - 1) = \kappa(t - s).$$

Let $0 < c < 1$. Then $d(M_c, M_1) = \kappa(1 - c) \leq \kappa$. Let σ be a unit speed geodesic such that

$$\sigma(0) \in M_1, \quad \dot{\sigma}(0) \perp M_1, \quad \text{and} \quad g(\dot{\sigma}(0), \nabla \mu_{\mathcal{R}}) = -1.$$

By Lemma 10.16 $d(\sigma(t), M_1) = t$. Consequently, $t < \kappa$ and the geodesic σ does not extend for infinite time. This shows $\mathcal{M}$ is incomplete and completes the proof of Theorem 10.17. □

10.4 3-DIMENSIONAL WALKER LORENTZIAN MANIFOLDS I: CURVATURE HOMOGENEITY

The material of this section and the next section is based on work of García-Río, Gilkey and Nikčević [67]. We will be concerned with the following examples.

Definition 10.19

1. The manifold $(\mathbb{R}^3, ds^2 := -2cy^2 dx \otimes dx + dy \otimes dy + dx \otimes d\tilde{x} + d\tilde{x} \otimes dx)$ for $c \in \mathbb{R}$ is called a *Cahen–Wallach space* and we refer to Cahen et al. [28] for further details. It is a 3-dimensional Lorentzian symmetric space and plays a central role in the theory. We postpone until Definition 11.23 a discussion of the higher-dimensional analogues that will play a crucial role in our analysis of Ricci solitons.
2. A 3-dimensional Lorentzian manifold $\mathcal{M} = (M, g_M)$ is said to be a *Walker manifold* if it admits a parallel null vector field. Such a manifold admits local coordinates of the following form. Let $f(x, y)$ be a smooth function on an open subset $\mathcal{O}$ of $\mathbb{R}^2$; we will usually assume that $\mathcal{O} = \mathbb{R}^2$ but that is not necessary. Let $\mathcal{M}_f := (\mathcal{O} \times \mathbb{R}, g_f)$ where $g_f(\partial_x, \partial_x) := -2f(x, y)$ and $g_f(\partial_x, \partial_{\tilde{x}}) = g_f(\partial_y, \partial_y) = 1$. The Christoffel symbols of such a manifold take the form:

$$\nabla_{\partial_x} \partial_x = -f_x \partial_{\tilde{x}} + f_y \partial_y, \quad \nabla_{\partial_x} \partial_y = \nabla_{\partial_y} \partial_x = -f_y \partial_{\tilde{x}}, \quad \nabla_{\partial_y} \partial_y = 0. \qquad (10.4.a)$$

   Thus, $\mathcal{M}_f$ is not a generalized plane wave manifold as defined in Equation (9.5.a).

3. Note that a Cahen–Wallach space is defined by taking $f = cy^2$ in Statement 2.

In this section, we determine which Lorentzian Walker manifolds are 0-curvature homogeneous (see Theorem 10.23), which are 1-curvature homogeneous (see Theorem 10.24), and which are 2-curvature homogeneous (see Theorem 10.25). We will show (see Theorem 10.26) that 2-curvature homogeneity is equivalent to local homogeneity in this context. In the next section, we will perform a similar investigation of homothety curvature homogeneity. We begin with some rather elementary results.

Lemma 10.20 Let $\mathcal{M}_f$ be as given in Definition 10.19.

1. The possibly non-zero curvatures of $\mathcal{M}_f$ are given by:
   (a) $R(\partial_x, \partial_y, \partial_y, \partial_x) = f_{yy}$.
   (b) $\nabla R(\partial_x, \partial_y, \partial_y, \partial_x; \partial_x) = f_{xyy}$, $\nabla R(\partial_x, \partial_y, \partial_y, \partial_x; \partial_y) = f_{yyy}$.
   (c) $\nabla^2 R(\partial_x, \partial_y, \partial_y, \partial_x; \partial_x, \partial_x) = f_{xxyy} - f_y f_{yyy}$, $\nabla^2 R(\partial_x, \partial_y, \partial_y, \partial_x; \partial_y, \partial_y) = f_{yyyy}$, $\nabla^2 R(\partial_x, \partial_y, \partial_y, \partial_x; \partial_x, \partial_y) = \nabla^2 R(\partial_x, \partial_y, \partial_y, \partial_x; \partial_y, \partial_x) = f_{xyyy}$.
2. $\mathcal{M}_f$ is a VSI manifold.
3. If $f(y) = cy^2$, then $\mathcal{M}_f$ is a symmetric space and is geodesically complete.
4. If $f(y) = -y^n$ for $n = 2, 3, \ldots$, then $\mathcal{M}_f$ is not geodesically complete and exhibits Ricci blowup; it cannot be embedded in a geodesically complete manifold.

Proof. We omit the proof of Assertion 1 as it is a straightforward computation. We prove Assertion 2 as follows. We use Equation (10.4.a) to see that range$\{\mathcal{R}\} \subset \operatorname{span}\{\partial_y, \partial_{\tilde{x}}\}$. Covariantly differentiating this relationship implies similarly that range$\{\nabla^k \mathcal{R}\} \subset \operatorname{span}\{\partial_y, \partial_{\tilde{x}}\}$ for any k. Furthermore, $\nabla^k \mathcal{R}(\cdot)$ vanishes if any entry is $\partial_{\tilde{x}}$ since $\nabla_{\tilde{x}} = 0$ and since the metric is independent of $\tilde{x}$. Lowering indices then shows that $\nabla^k R(\cdot) = 0$ if any index is $\partial_{\tilde{x}}$. Thus, the only non-zero entries take the form $\nabla^k R(\partial_x, \partial_y, \partial_y, \partial_x; \ldots)$. Relative to the basis $\{\partial_x, \partial_{\tilde{x}}, \partial_y\}$, the metric tensor g_{ij} and the inverse metric tensor g^{ij} are given by:

$$(g_{ij}) = \begin{pmatrix} -2f & 1 & 0 \\ 1 & 0 & 0 \\ 0 & 0 & 1 \end{pmatrix} \quad \text{and} \quad (g^{ij}) = \begin{pmatrix} 0 & 1 & 0 \\ 1 & 2f & 0 \\ 0 & 0 & 1 \end{pmatrix}.$$

Thus, any Weyl contraction must involve a $\partial_{\tilde{x}}$ variable; the curvature tensor vanishes on such variables. Consequently, these manifolds are VSI. Assertion 2 now follows.

We now prove Assertion 3 and Assertion 4; Assertion 3 provides examples of VSI manifolds which are not generalized plane wave manifolds. Let $x^1 = x$, $x^2 = y$, and $x^3 = \tilde{x}$. By Equation (10.4.a), the (possibly) non-zero Christoffel symbols are given by $\Gamma_{11}{}^3 = -f_x$, $\Gamma_{11}{}^2 = f_y$, and $\Gamma_{12}{}^3 = \Gamma_{21}{}^3 = -f_y$. The geodesic equation $\ddot{x}^k + \Gamma_{ij}{}^k \dot{x}^i \dot{x}^j = 0$ becomes

$$\ddot{x}^1 = 0, \quad \ddot{x}^2 = -\dot{x}^1 \dot{x}^1 f_y, \quad \ddot{x}^3 = \dot{x}^1 \dot{x}^1 f_x + 2\dot{x}^1 \dot{x}^2 f_y.$$

The first equation implies that $x^1(t) = x_0^1 + at$ is linear. If the second equation can be solved then x^3 is determined. Thus, the crucial equation is $\ddot{x}^2 = -a^2 f_y$. If $f(y) = cy^2$, this equation becomes $\ddot{x}^2 = -2a^2cx^2$; the solution to this are either exponentials (if $-2a^2c > 0$) or sine and cosine functions (if $-2a^2c < 0$). Assertion 3 now follows. Suppose

$$f(y) = -y^n \quad \text{so} \quad f_y = -ny^{n-1}\,.$$

The geodesic equation is $\ddot{x}^2(t) = a^2n(x^2)^n$. Let $x^2(t) = (2-t)^\alpha$ for $\alpha < 0$. We require

$$\alpha(\alpha-1)(2-t)^{\alpha-2} = a^2n(2-t)^{\alpha(n-1)}\,.$$

Given $\alpha < 0$, we choose a so $\alpha(\alpha-1) = a^2n$ and require $\alpha - 2 = (n-1)\alpha$ so $\alpha = \frac{2}{2-n}$. This is negative since $n > 2$. Since $\lim_{t\to 2} x^2(t) = \infty$, the manifold in question is geodesically incomplete. The Ricci tensor $\rho = f_{yy}dx \otimes dx$ takes the form $\rho = n(n-1)(x^2)^{n-2}$. Thus, $\lim_{t\to 2}\rho(\dot{\gamma},\dot{\gamma}) = \infty$. This shows $\mathcal{M}_f$ exhibits Ricci blowup. Consequently, $\mathcal{M}$ cannot be embedded in a geodesically complete manifold. □

Definition 10.21 A tensor T is said to be *recurrent* if there is a smooth 1-form ω so that $\nabla_X T = \omega(X)T$. An affine manifold $\mathcal{M} = (M, \nabla)$ is said to be *recurrent* if the curvature of ∇ is recurrent.

If $f_{yy} \neq 0$ and if $\nabla R \neq 0$, then Assertion 1 of Lemma 10.20 shows the manifolds $\mathcal{M}_f$ are recurrent where we take $\omega = \frac{f_{xyy}}{f_{yy}}dx + \frac{f_{yyy}}{f_{yy}}dy$.

10.4.1 RENORMALIZING THE COORDINATE SYSTEM.

Lemma 10.22 Let $T(x, y, \tilde{x}) = (x, y+\phi, \tilde{x} - \phi_x y + \psi)$ where ϕ and ψ are smooth functions of x. Then T is an isometry from $\mathcal{M}_f$ to $\mathcal{M}_{\tilde{f}}$ for $\tilde{f} = f(x, y+\phi) + \phi_{xx}y - \psi_x - \frac{1}{2}\phi_x^2$.

Proof. We show that $T^*(g_f) = g_{\tilde{f}}$ by computing: $T_*\partial_x = \partial_x + \phi_x\partial_y + (-\phi_{xx}y + \psi_x)\partial_{\tilde{x}}$, $T_*\partial_y = \partial_y - \phi_x\partial_{\tilde{x}}$, and $T_*\partial_{\tilde{x}} = \partial_{\tilde{x}}$. Thus,

$$\begin{aligned}
&g_f(T_*\partial_x, T_*\partial_x) = -2\{f(x, y+\phi) + \phi_{xx}y - \psi_x - \tfrac{1}{2}\phi_x^2\},\\
&g_f(T_*\partial_x, T_*\partial_{\tilde{x}}) = 1, \qquad g_f(T_*\partial_y, T_*\partial_y) = 1, \qquad g_f(T_*\partial_x, T_*\partial_y) = -\phi_x + \phi_x = 0,\\
&g_f(T_*\partial_y, T_*\partial_{\tilde{x}}) = g_f(T_*\partial_{\tilde{x}}, T_*\partial_{\tilde{x}}) = 0\,.
\end{aligned}$$

□

10.4.2 0-CURVATURE HOMOGENEITY. Since $R(\partial_x, \partial_y, \partial_y, \partial_x) = f_{yy}$, it is natural to assume f_{yy} is never zero. We will often suppose that $f_{yy} > 0$ henceforth as the case $f_{yy} < 0$ is analogous.

Theorem 10.23 *If $f_{yy} > 0$, then $\mathcal{M}_f$ is 0-curvature homogeneous.*

Proof. We have $\ker\{\mathcal{R}\} = \operatorname{span}\{\partial_{\tilde{x}}\}$ and $\operatorname{range}\{\mathcal{R}\} = \operatorname{span}\{\partial_{\tilde{x}}, \partial_y\}$. Thus, these two subspaces are invariantly defined. We set

$$\xi_1 := a_{11}(\partial_x + f\partial_{\tilde{x}} + a_{12}\partial_y + a_{13}\partial_{\tilde{x}}), \quad \xi_2 := a_{22}\partial_y + a_{23}\partial_{\tilde{x}}, \quad \xi_3 := a_{33}\partial_{\tilde{x}}\,.$$

These form a pseudo-orthonormal basis. We wish to ensure that

$$\langle \xi_1, \xi_3 \rangle = \langle \xi_2, \xi_2 \rangle = 1 \quad \text{and} \quad R(\xi_1, \xi_2, \xi_2, \xi_1) = 1\,. \tag{10.4.b}$$

This gives rise to the equations $2a_{13} + a_{12}^2 = 0$, $a_{23} + a_{12}a_{22} = 0$, $a_{22}^2 = 1$, $a_{11}a_{33} = 1$, and $a_{11}^2 a_{22}^2 f_{yy} = 1$. Let a_{12} be arbitrary for the moment. We have

$$a_{11} = f_{yy}^{-1/2}, \; a_{12} = \star, \; a_{13} = -\tfrac{1}{2}a_{12}^2, \; a_{22} = 1, \; a_{23} = -a_{12}, \; a_{33} = f_{yy}^{1/2}\,. \tag{10.4.c}$$

The parameters a_{13}, a_{23}, and a_{33} play no further role. Since Equation (10.4.b) is satisfied, $\mathcal{M}_f$ is 0-curvature homogeneous. □

10.4.3 1-CURVATURE HOMOGENEITY.

Theorem 10.24 *Assume that $f_{yy} > 0$. Then $\mathcal{M}_f$ is 1-curvature homogeneous if and only if exactly one of the following possibilities holds.*

1. *$f_{yy}(x, y) = \alpha(x)e^{by}$ where $0 \neq b \in \mathbb{R}$ and where $\alpha(x)$ is arbitrary.*
2. *$f_{yy}(x, y) = c(x - x_0)^{-2}$ for some $x_0 \in \mathbb{R}$ and some $0 \neq c \in \mathbb{R}$.*
3. *f_{yy} is constant.*

Proof. We will adopt the normalizations of Equation (10.4.c) to ensure that Equation (10.4.b) is satisfied. The parameter a_{12} is still a free parameter. We compute that

$$\nabla R(\xi_1, \xi_2, \xi_2, \xi_1; \xi_2) = f_{yyy} \cdot f_{yy}^{-1}\,.$$

Since a_{12} plays no role, $f_{yyy} \cdot f_{yy}^{-1}$ is an isometry invariant. Consequently, if $\mathcal{M}_f$ is 1-curvature homogeneous, then $f_{yyy} = b \cdot f_{yy}$ for some $b \in \mathbb{R}$ and thus $f_{yyy} = \alpha(x)e^{by}$. The possibility in Assertion 1 arises from $b \neq 0$ and the possibility in Assertion 2 arises from $b = 0$ and $\alpha(x)$ non-constant.

Case 1. If $b \neq 0$, then $f_{yyy} \neq 0$ and we may set:

$$\begin{aligned} a_{11} &= f_{yy}^{-1/2}, & a_{12} &= -f_{xyy} \cdot f_{yyy}^{-1}, & a_{13} &= -\tfrac{1}{2}a_{12}^2, \\ a_{22} &= 1, & a_{23} &= -a_{12}, & a_{33} &= f_{yy}^{1/2}. \end{aligned} \tag{10.4.d}$$

With these normalizations, we show $\mathcal{M}_f$ is 1-curvature homogeneous and establish Assertion 1 by computing:

$$\begin{aligned} R(\xi_1,\xi_2,\xi_2,\xi_1) &= f_{yy} \cdot f_{yy}^{-1} = 1, \\ \nabla R(\xi_1,\xi_2,\xi_2,\xi_1;\xi_1) &= \{f_{xyy} + a_{12} f_{yyy}\} f_{yy}^{-3/2} = 0, \\ \nabla R(\xi_1,\xi_2,\xi_2,\xi_1;\xi_2) &= f_{yyy} \cdot f_{yy}^{-1} = b\,. \end{aligned} \tag{10.4.e}$$

Case 2. If $b = 0$, then a_{12} plays no role in the computation of ∇R so:

$$\begin{aligned} R(\xi_1,\xi_2,\xi_2,\xi_1) &= f_{yy} \cdot f_{yy}^{-1} = 1, \\ \nabla R(\xi_1,\xi_2,\xi_2,\xi_1;\xi_1) &= f_{xyy} \cdot f_{yy}^{-3/2} = \alpha_x \cdot \alpha^{-3/2}, \\ \nabla R(\xi_1,\xi_2,\xi_2,\xi_1;\xi_2) &= f_{yyy} \cdot f_{yy}^{-1} = 0\,. \end{aligned}$$

So $\mathcal{M}_f$ will be 1-curvature homogeneous if and only if $\alpha_x = c_{12211}\alpha^{3/2}$ for some non-zero constant c_{12211}. We solve this ODE to see $\alpha = c(x - x_0)^{-2}$.

Case 3. f_{yy} is constant. In this instance, $\mathcal{M}_f$ is a symmetric space and, therefore, k-curvature homogeneous for all k. □

10.4.4 2-CURVATURE HOMOGENEITY.

Theorem 10.25 *The manifold $\mathcal{M}_f$ is 2-curvature homogeneous if and only if it falls into one of the three families.*

1. $f = b^{-2}\alpha(x)e^{by} + \beta(x)y + \gamma(x)$ *for* $\beta(x) = b^{-1}\alpha^{-1}\{\alpha_{xx} - \alpha_x^2\alpha^{-1}\}$, $b \neq 0$ *and* $\alpha > 0$.
2. $f_{yy} = \alpha(x) > 0$ *where* $\alpha_x = c\alpha^{3/2}$ *for* $c > 0$. *Consequently,* $\alpha = \tilde{c}(x - x_0)^{-2}$ *for some* $(\tilde{c}, x_0)$. *This means* $f = a(x - x_0)^{-2}y^2 + \beta(x)y + \gamma(x)$ *where* $0 \neq a \in \mathbb{R}$.
3. $f = \varepsilon y^2 + \beta(x)y + \gamma(x)$ *where* $0 < \varepsilon \in \mathbb{R}$.

Proof. We adopt the normalizations of Equation (10.4.d) and continue the computations of Equation (10.4.e) to see

$$\begin{aligned} \nabla^2 R(\xi_1,\xi_2,\xi_2,\xi_1;\xi_1,\xi_2) &= f_{yy}^{-3/2}\{f_{xyyy} - f_{yyyy} f_{xyy}/f_{yyy}\} \\ &= f_{yy}^{-3/2}\{b\alpha_x - b\alpha_x\}e^{by} = 0, \\ \nabla^2 R(\xi_1,\xi_2,\xi_2,\xi_1;\xi_2,\xi_2) &= f_{yy}^{-1} f_{yyyy} = b^2, \\ \nabla^2 R(\xi_1,\xi_2,\xi_2,\xi_1;\xi_1,\xi_1) &= f_{yy}^{-2}\{\nabla^2 R(\partial_x,\partial_y,\partial_y,\partial_x;\partial_x,\partial_x) \\ &\quad + 2a_{12}\nabla^2 R(\partial_x,\partial_y,\partial_y,\partial_x;\partial_x,\partial_y) + a_{12}^2\nabla^2 R(\partial_x,\partial_y,\partial_y,\partial_x;\partial_y,\partial_y)\}\,. \end{aligned}$$

Thus, only $\nabla^2 R(\xi_1, \xi_2, \xi_2, \xi_1; \xi_1, \xi_1)$ is relevant to our discussion. We assume that $\mathcal{M}_f$ is 1-curvature homogeneous and examine the cases of Theorem 10.24 seriatim.

Case 1. Let $f = b^{-2}\alpha(x)e^{by} + \beta(x)y + \gamma(x)$. We set $a_{12} = -f_{xyy} \cdot f_{yyy}^{-1}$ and expand

$$\begin{aligned}\nabla^2 R(\xi_1, \xi_2, \xi_2, \xi_1; \xi_1, \xi_1) &= f_{yy}^{-2}\{\nabla^2 R(\partial_x, \partial_y, \partial_y, \partial_x; \partial_x, \partial_x)\\ &\quad +2a_{12}\nabla^2 R(\partial_x, \partial_y, \partial_y, \partial_x; \partial_x, \partial_y) + a_{12}^2\nabla^2 R(\partial_x, \partial_y, \partial_y, \partial_x; \partial_y, \partial_y)\}\\ &= f_{yy}^{-2}\{f_{xxyy} - f_y f_{yyy} - 2f_{xyy}f_{yyy}^{-1}f_{xyyy} + f_{xyy}^2 f_{yyy}^{-2} f_{yyyy}\}\\ &= e^{-by}\alpha^{-2}\{\alpha_{xx} - b\alpha\beta(x) - 2\alpha_x^2\alpha^{-1} + \alpha_x^2\alpha^{-1}\} - 1\,.\end{aligned}$$

We complete the proof in this special case by setting $\beta(x) = b^{-1}\alpha^{-1}\{\alpha_{xx} - \alpha_x^2\alpha^{-1}\}$ to see that $\mathcal{M}_f$ is 2-curvature homogeneous.

Case 2. Suppose $f_{yy} = \alpha(x) > 0$ where $\alpha_x = c\alpha^{3/2}$ for $0 \neq c \in \mathbb{R}$. We adopt the normalizations of Equation (10.4.c); the parameter a_{12} plays no role. We compute:

$$\begin{aligned}\nabla^2 R(\xi_1, \xi_2, \xi_2, \xi_1; \xi_1, \xi_2) &= \alpha^{-3/2} f_{xyyy} = 0,\\ \nabla^2 R(\xi_1, \xi_2, \xi_2, \xi_1; \xi_2, \xi_2) &= \alpha^{-1} f_{yyyy} = 0\,.\end{aligned}$$

Thus, only $\nabla^2 R(\xi_1, \xi_2, \xi_2, \xi_1; \xi_1, \xi_1)$ is relevant. The f_{yyy} term no longer plays a role so

$$\nabla^2 R(\xi_1, \xi_2, \xi_2, \xi_1; \xi_1, \xi_1) = \alpha^{-2}\alpha_{xx}\,.$$

We have $\alpha_x = c\alpha^{3/2}$ and thus $\alpha_{xx} = \frac{3}{2}c \cdot \alpha_x \cdot \alpha^{1/2} = \frac{3}{2}c^2\alpha^2$, from where it follows that $\nabla^2 R(\xi_1, \xi_2, \xi_2, \xi_1; \xi_1, \xi_1)$ is constant. Hence, $\mathcal{M}_f$ is 2-curvature homogeneous.

We proceed inductively to show that the only non-zero entry in the k-th covariant derivative $\nabla^k R$ is given by $\nabla^k R(\xi_1, \xi_2, \xi_2, \xi_1; \xi_1, \dots, \xi_1)$ and that

$$\nabla^{k-1} R(\partial_x, \partial_y, \partial_y, \partial_x; \partial_x, \dots, \partial_x) = c_{k-1}\alpha^{(1+k)/2}\,.$$

It then follows that

$$\begin{aligned}\nabla^k R(\partial_x, \partial_y, \partial_y, \partial_x; \partial_x, \dots, \partial_x) &= c_{k-1}\tfrac{1+k}{2}\alpha_x\alpha^{(-1+k)/2}\\ &= c_{k-1}c\tfrac{1+k}{2}\alpha^{(3-1+k)/2} = c_k\alpha^{(2+k)/2} \quad \text{for} \quad c_k := c_{k-1}c\tfrac{1+k}{2},\\ \nabla^k R(\xi_1, \xi_2, \xi_2, \xi_1; \xi_1, \dots, \xi_1) &= \alpha^{(-2-k)/2}\nabla^k R(\partial_x, \partial_y, \partial_y; \partial_x, \dots, \partial_x) = c_k\,.\end{aligned}$$

Thus, $\mathcal{M}_f$ is k-curvature homogeneous for all k and, consequently, locally homogeneous.

Case 3. f_{yy} is constant. In this instance, $\mathcal{M}_f$ is a symmetric space, k-curvature homogeneous for all k, and locally homogeneous. □

10.4.5 LOCAL HOMOGENEITY.

Theorem 10.26 *Suppose that $\mathcal{M}_f$ is 2-curvature homogeneous. Then $\mathcal{M}_f$ is locally isometric to an element of one of the following three families, all of which are homogeneous.*

1. $f = b^{-2}e^{by}$.
2. $f = a(x-x_0)^{-2}y^2$ *for* $0 \neq a$.
3. $f = \varepsilon y^2$ *for* $\varepsilon > 0$.

We note that these three families have different 2-curvature models and are thus non-isomorphic. If $f = \varepsilon y^2$, then (see Definition 10.19), $\mathcal{M}_f$ is a *Cahen–Wallach space.*

Proof. We apply Lemma 10.22 to examine the three cases of Theorem 10.25 seriatim.

Case 1. Let $f = b^{-2}\alpha(x)e^{by} + b^{-1}\alpha^{-1}\{\alpha_{xx} - \alpha_x^2\alpha^{-1}\}y + \gamma(x)$ for $\alpha > 0$. Let $F = b^{-2}e^{by}$. Set $\phi = \ln(\alpha)b^{-1}$. Choose ψ so that $-\psi_x - \frac{1}{2}\phi_x^2 = \gamma$. We compute:

$$\begin{aligned}\tilde{f}(x,y) &= b^{-2}e^{b(y+b^{-1}\ln(\alpha))} + b^{-1}\{\alpha^{-1}\alpha_x\}_x y + \gamma \\ &= b^{-2}\alpha e^{by} + b^{-1}\alpha^{-1}\{\alpha_{xx} - \alpha_x^2\alpha^{-1}\}y + \gamma = f\,.\end{aligned}$$

This shows we may replace f by $b^{-2}e^{by}$. Suppose given a point $(a_1,a_2,a_3) \in \mathbb{R}^3$. We consider the map $T(x,y,\tilde{x}) = (e^{-ba_2/2}x + a_1, y + a_2, e^{ba_2/2}\tilde{x} + a_3)$. This family of transformations acts transitively on $\mathcal{M}_{b^{-2}e^{by}}$ by isometries and thus $\mathcal{M}_{b^{-2}e^{by}}$ is a homogeneous space.

Case 2. Suppose $f_{yy} = \tilde{c}(x-x_0)^{-2}$. Assume that $f = \frac{1}{2}y^2\alpha_c$. Suppose that $\beta = \beta(x)$ and $\gamma = \gamma(x)$ are given. Choose ϕ so $\alpha_c\phi + \phi_{xx} = \beta$ and ψ so $-\psi_x - \frac{1}{2}\phi_x^2 + \frac{1}{2}\alpha_c\phi = \alpha$. One then has that

$$\begin{aligned}\tilde{f}(x,y) &= \tfrac{1}{2}y^2\alpha_c(x) + y(\alpha_c\phi + \phi_{xx}) - \psi_x - \tfrac{1}{2}\phi_x^2 + \tfrac{1}{2}\alpha_c\phi^2 \\ &= \tfrac{1}{2}y^2\alpha_c(x) + y\beta + \gamma\,.\end{aligned}$$

This shows that we may replace f by $f(x,y) = cy^2(x+1)^{-2}$ and M by $(-1,\infty)\times\mathbb{R}^2$; the case when $f(x,y) = cy^2(x-1)^{-2}$ and $M = (-\infty,1)\times\mathbb{R}^2$ is handled similarly (the question of where the singularity is relative to $x=0$ plays an important role). Let $(a_1,a_2,a_3)\in\mathbb{R}^3$ be given with $a_1 > -1$. Choose ϕ and ψ so that

$$\begin{aligned} &c(x+1)^{-2}2\phi + (a_1+1)\phi_{xx} = 0, &&\phi(0) = a_2, \\ &c(x+1)^{-2}\phi^2 - \tfrac{1}{2}\phi_x^2 - (a_1+1)\psi_x = 0, &&\psi(0) = a_3\,.\end{aligned}$$

Set $T(x,y,\tilde{x}) = ((a_1+1)x + a_1, y + \phi, (a_1+1)^{-1}\tilde{x} - \phi_x y + \psi)$. We show that this family provides a transitive action by isometries and that therefore $\mathcal{M}_{cy^2(x+1)^{-2}}$ is a homogeneous space by computing:

$$T_*\partial_x = (a_1+1)\partial_x + \phi_x\partial_y + (-\phi_{xx}y + \psi_x)\partial_{\tilde{x}}, \quad T_*\partial_y = \partial_y - \phi_x\partial_{\tilde{x}},$$

$$T_*\partial_{\tilde{x}} = (a_1+1)^{-1}\partial_{\tilde{x}}, \quad g_f(T_*\partial_x, T_*\partial_{\tilde{x}}) = 1, \quad g_f(T_*\partial_y, T_*\partial_y) = 1,$$

$$g_f(T_*\partial_x, T_*\partial_y) = 0, \quad g_f(T_*\partial_y, T_*\partial_{\tilde{x}}) = 0, \quad g_f(T_*\partial_{\tilde{x}}, T_*\partial_{\tilde{x}}) = 0,$$

$$g_f(T_*\partial_x, T_*\partial_x) = -2\left\{(a_1+1)^2 c((a_1+1)x + a_1 + 1)^{-2}(y^2 + 2\phi y + \phi^2) \right. \\ \left. -\tfrac{1}{2}\phi_x^2 + (a_1+1)(\phi_{xx}y - \psi_x)\right\} = f\,.$$

Case 3. Let $f(x, y) = \varepsilon y^2$ where $\varepsilon > 0$. Let $\beta = \beta(x)$ and $\gamma = \gamma(x)$ be given. Choose ϕ so that $2\varepsilon\phi + \phi_{xx} = \beta$; ϕ need not be globally defined, but this is always possible locally. Then choose ψ so $-\psi_x - \frac{1}{2}\phi_x^2 + \varepsilon\phi^2 = \gamma$. We then have that

$$\tilde{f}(x, y) = \varepsilon y^2 + (2\varepsilon\phi + \phi_{xx})y - \psi_x - \tfrac{1}{2}\phi_x^2 + \varepsilon\phi^2 = \varepsilon y^2 + \beta y + \gamma\,.$$

Consequently, T is a local isometry between $\mathcal{M}_{\varepsilon y^2}$ and $\mathcal{M}_{\varepsilon y^2+\beta y+\gamma}$. Since the transformation $T_\varepsilon(x, y, \tilde{x}) = (\sqrt{\varepsilon}x, y, \tilde{x}/\sqrt{\varepsilon})$ provides an isometry between $\mathcal{M}_{y^2}$ and $\mathcal{M}_{\varepsilon y^2}$, the parameter ε plays no role. Suppose given a point $(a_1, a_2, a_3) \in \mathbb{R}$. Set $\phi(x) = a_2\cos(\sqrt{2}x)$. We then have that $2\phi + \phi_{xx} = 0$ and $\phi(0) = a_2$. Now choose $\psi(x)$ so that $\psi_x + \frac{1}{2}\phi_x^2 + \phi^2 = 0$ and so that $\psi(0) = a_3$. Let $T(x, y, z) = (x + a_1, y + \phi, \tilde{x} - \phi_x y + \psi)$. The translation in the x coordinate is harmless and does not change the equations of structure. We then have that $T^*g_f = g_f$ and $T(0, 0, 0) = (a_1, a_2, a_3)$. Consequently, $\mathcal{M}_{y^2}$ is globally a homogeneous space. □

10.5 WALKER LORENTZIAN MANIFOLDS II: HOMOTHETY CURVATURE HOMOGENEITY

We extend the results of Section 10.4 from the setting of curvature homogeneity to the setting of homothety curvature homogeneity. We computed R and ∇R in Lemma 10.20. We have that $R(\partial_x, \partial_y, \partial_y, \partial_x) = f_{yy}$. The vanishing of f_{yy} is an invariant of the homothety 0-model. Since we are interested in homothety curvature homogeneity, we will assume f_{yy} never vanishes; as before, we will assume $f_{yy} > 0$. By Theorem 10.23, this implies that $\mathcal{M}_f$ is 0-curvature homogeneous and, consequently, 0-homothety curvature homogeneous as well. The first interesting case, therefore, is that of 1-homothety curvature homogeneity. This will be dealt with in Theorem 10.27; we will, of course, be interested in examples which are not covered by Theorem 10.24.

10.5.1 1-HOMOTHETY CURVATURE HOMOGENEITY. Lemma 10.20 yields

$$\nabla R(\partial_x, \partial_y, \partial_y, \partial_x; \partial_x) = f_{xyy} \quad \text{and} \quad \nabla R(\partial_x, \partial_y, \partial_y, \partial_x; \partial_y) = f_{yyy}\,.$$

The simultaneous vanishing of f_{yyy} and f_{xyy} is an invariant of the homothety 1-model. The case $f_{yy} = a$ for $0 \neq a \in \mathbb{R}$ gives rise to a Cahen–Wallach space (see Definition 10.19). We will therefore assume f_{yy} non-constant. This gives rise to two cases. In the first, f_{yyy} is never zero and in the second, f_{yyy} vanishes identically but f_{xyy} is never zero.

Theorem 10.27 *Suppose $f_{yy} > 0$ is never zero and non-constant.*

1. *If f_{yyy} never vanishes, then $\mathcal{M}_f$ is homothety 1-curvature homogeneous.*
2. *If $f_{yy} = \alpha(x)$ with α_x never zero, then $\mathcal{M}_f$ is homothety 1-curvature homogeneous if and only if $f = a(x-x_0)^{-2}y^2 + \beta(x)y + \gamma(x)$ where $0 \neq a \in \mathbb{R}$. By Theorem 10.26, this manifold is locally homogeneous.*

Proof. We suppose $f_{yy} > 0$ as the case $f_{yy} < 0$ is analogous. The two distributions

$$\ker\{\mathcal{R}\} = \operatorname{span}\{\partial_{\tilde{x}}\} \quad \text{and} \quad \operatorname{range}\{\mathcal{R}\} = \operatorname{span}\{\partial_y, \partial_{\tilde{x}}\}$$

are invariantly defined. To preserve these distributions, we set:

$$\xi_1 = a_{11}(\partial_x + f\partial_{\tilde{x}} + a_{12}\partial_y + a_{13}\partial_{\tilde{x}}), \quad \xi_2 = \partial_y + a_{23}\partial_{\tilde{x}}, \quad \xi_3 = a_{33}\partial_{\tilde{x}}, \tag{10.5.a}$$

for some functions a_{ij} on $\mathcal{O}$. To ensure that the inner products are normalized properly, we impose the relations:

$$a_{12}^2 + 2a_{13} = 0, \quad a_{12} + a_{23} = 0, \quad a_{11}a_{33} = 1.$$

This determines a_{13}, a_{23}, and a_{33}; these parameters play no further role and $\{\lambda, a_{11}, a_{12}\}$ remain as free parameters where λ is the homothety rescaling factor. If $f_{yyy} \neq 0$, set

$$\lambda := f_{yyy}f_{yy}^{-1}, \quad a_{12} := -f_{xyy}f_{yyy}^{-1}, \quad a_{11}^2 := f_{yy}^{-1}\lambda^2. \tag{10.5.b}$$

We then have

$$\begin{aligned} R(\xi_1,\xi_2,\xi_2,\xi_1) &= a_{11}^2 f_{yy} = \lambda^2, \\ \nabla R(\xi_1,\xi_2,\xi_2,\xi_1;\xi_1) &= a_{11}^3\{f_{xyy} + a_{12}f_{yyy}\} = 0, \\ \nabla R(\xi_1,\xi_2,\xi_2,\xi_1;\xi_2) &= a_{11}^2 f_{yyy} = \lambda^2 f_{yy}^{-1} f_{yyy} = \lambda^3. \end{aligned} \tag{10.5.c}$$

All the parameters have been determined (modulo a possible sign ambiguity in a_{11}) and it follows $\mathcal{M}_f$ is 1-homothety curvature homogeneous. This proves Assertion 1.

We now prove Assertion 2. Suppose $f_{yy} > 0$, $f_{yyy} = 0$, and f_{xyy} never vanishes. Set $f_{yy} = \alpha(x)$. The parameter a_{12} plays no role. To ensure that $\mathcal{M}_f$ is homothety 1-curvature homogeneous, we impose the following relations where $\{a_{11}, \lambda\}$ are unknown functions to be determined and where $\{c_0, c_1\}$ are unknown constants:

$$\begin{aligned} R(\xi_1,\xi_2,\xi_2,\xi_1) &= a_{11}^2(x)\alpha(x) = \lambda^2(x)c_0, \\ R(\xi_1,\xi_2,\xi_2,\xi_1;\xi_1) &= a_{11}^3(x)\alpha_x(x) = \lambda^3(x)c_1, \\ R(\xi_1,\xi_2,\xi_2,\xi_1;\xi_2) &= 0. \end{aligned}$$

Thus, $a_{11}^6(x)\alpha^3(x) = \lambda^6(x)c_0^3$ and $a_{11}^6(x)\alpha_x^2(x) = \lambda^6(x)c_1^2$. This shows that $\alpha^3(x) = c_3\alpha_x^2(x)$ for some constant c_3. We solve this ordinary differential equation to complete the proof by checking $\alpha(x) = a(x-x_0)^{-2}$ for $0 \neq a \in \mathbb{R}$ and $x_0 \in \mathbb{R}$. □

10.5.2 2-HOMOTHETY CURVATURE HOMOGENEITY. We continue the analysis of the manifolds in Assertion 1 of Theorem 10.27 as these are the only possible source of new examples not covered by Theorem 10.25.

Theorem 10.28 *Assume that $\mathcal{M}_f$ is homothety 2-curvature homogeneous, and that f_{yy} and f_{yyy} never vanish. Then $\mathcal{M}_f$ is locally isometric to one of the following examples.*

1. *$f = \pm e^{ay}$ for some $a \neq 0$ and $M = \mathbb{R}^3$. $\mathcal{M}_f$ is homogeneous.*
2. *$f = \pm \ln(y)$ and $M = \mathbb{R} \times (0, \infty) \times \mathbb{R}$. $\mathcal{M}_f$ is homothety homogeneous, not locally homogeneous, and of cohomogeneity one.*
3. *$f = \pm y^{\varepsilon}$ for $\varepsilon \neq 0, 1, 2$ and $M = \mathbb{R} \times (0, \infty) \times \mathbb{R}$. $\mathcal{M}$ is homothety homogeneous, not locally homogeneous, and of cohomogeneity one.*

Proof. We will suppose $f_{yy} > 0$; the case $f_{yy} < 0$ is handled similarly. As any two homothety 1-curvature models for $\mathcal{M}_f$ are isomorphic, we can adopt the normalizations of Equation (10.5.a), (10.5.b), and (10.5.c). We have:

$$\begin{aligned}
&R(\xi_1, \xi_2, \xi_2, \xi_1) = a_{11}^2 f_{yy} = \lambda^2, \\
&\nabla R(\xi_1, \xi_2, \xi_2, \xi_1; \xi_2) = a_{11}^2 f_{yyy} = \lambda^3, \\
&\nabla^2 R(\xi_1, \xi_2, \xi_2, \xi_1; \xi_2, \xi_2) = a_{11}^2(x) f_{yyyy} = \lambda^4 c_{122122}, \\
&\frac{f_{yy} \cdot f_{yyyy}}{f_{yyy} \cdot f_{yyy}} = \frac{\lambda^2 a_{11}^{-2} \cdot \lambda^4 c_{11} a_{11}^{-2}}{\lambda^6 a_{11}^{-4}} = c_{122122} \,.
\end{aligned}$$

Thus, c_{122122} is an invariant of the theory; this will imply the three families of the theory fall into different local isometry types. The ordinary differential equation $\frac{\alpha\alpha''}{\alpha'\alpha'} = c_{122122}$ has the solutions (see, for example, Gilkey [73, Lemma 1.5.5]) of the form $\alpha(t) = e^{a(t+b)}$ or $\alpha(t) = a(t+b)^c$ where $a \neq 0$ and $c \neq 0$. Thus, there exists $\alpha(x) \neq 0$ and $c \neq 0$ so that

$$f_{yy} = e^{\alpha(x)(y+\beta(x))} \quad \text{or} \quad f_{yy} = \alpha(x)(y + \beta(x))^c \,. \tag{10.5.d}$$

Let $T(x, y, z) = (x, y - \beta(x), \tilde{x} + y\beta_x(x))$. Setting $\phi = -\beta(x)$ and $\psi = 0$ in Lemma 10.22 shows that $\mathcal{M}_f$ is isometric to $\mathcal{M}_{\tilde{f}}$ where

$$\tilde{f}(x, y) = f(x, y - \beta(x)) - \tfrac{1}{2}\left\{\beta_x^2(x) + 2y\beta_{xx}(x)\right\} \,.$$

Thus, we may assume henceforth that $\beta(x) = 0$ in Equation (10.5.d), i.e.,

$$f_{yy} = e^{\alpha(x)y} \quad \text{or} \quad f_{yy} = \alpha(x)y^c \,.$$

We examine these two cases seriatim. We will use the relations:

$$\begin{aligned}
\lambda &= f_{yyy} f_{yy}^{-1}, \quad a_{12} = -f_{xyy} f_{yyy}^{-1}, \quad \lambda^2 a_{11}^{-2} = f_{yy}, && (10.5.\text{e}) \\
\lambda^4 c_{122112} &= \nabla^2 R(\xi_1, \xi_2, \xi_2, \xi_1; \xi_1, \xi_2) = a_{11}^3\{f_{xyyy} + a_{12} f_{yyyy}\}, && (10.5.\text{f}) \\
\lambda^4 c_{122111} &= \nabla^2 R(\xi_1, \xi_2, \xi_2, \xi_1; \xi_1, \xi_1) && (10.5.\text{g}) \\
&= a_{11}^4\{f_{xxyy} + 2a_{12} f_{xyyy} + a_{12}^2 f_{yyyy} - f_y f_{yyy}\} \,.
\end{aligned}$$

Case I. Suppose $f_{yy} = e^{\alpha(x)y}$. Then Equation (10.5.e) implies that

$$\lambda = f_{yyy} f_{yy}^{-1} = \alpha(x), \; a_{12} = -f_{xyy} f_{yyy}^{-1} = -y\alpha_x(x)\alpha(x)^{-1}, \; \lambda^2 a_{11}^{-2} = f_{yy} = e^{\alpha(x)y}.$$

We use Equation (10.5.f) to see that:

$$\begin{aligned} f_{xyyy} + a_{12} f_{yyyy} &= \partial_x\{\alpha(x)e^{\alpha(x)y}\} - \alpha_x(x)\alpha(x)e^{\alpha(x)y} \\ &= \alpha_x(x) \cdot e^{\alpha(x)y} = a_{11}^{-3}\lambda^4 c_{12} = \alpha(x)e^{\frac{3}{2}\alpha(x)y} c_{122112}. \end{aligned}$$

It now follows that $\alpha_x(x) = 0$ so $\alpha(x) = a$ is constant and $f(x, y) = a^{-2}e^{ay} + u(x)y + v(x)$. We then use Equation (10.5.e) to see $\lambda = a$, $a_{12} = 0$, and $\lambda^2 a_{11}^{-2} = e^{ay}$. Equation (10.5.g) then leads to the identity:

$$e^{2ay} c_{122111} = a_{11}^{-4}\lambda^4 c_{122111} = -f_y f_{yyy} = -e^{2ay} - u(x)ae^{ay}.$$

This implies that $u(x) = 0$ and, therefore, $f = a^{-2}e^{ay} + v(x)$. We set $\phi = 0$ and choose ψ so $\psi(x) = v$ and apply Lemma 10.22 to see that $\mathcal{M}_f$ is isometric to $\mathcal{M}_{\tilde{f}}$ where $\tilde{f} := a^{-2}e^{ay}$. Replacing y by $y + y_0$ for suitably chosen y_0, then replaces f by e^{ay} as desired. It then follows by Theorem 10.26 that $\mathcal{M}_f$ is locally homogeneous.

Case II. Suppose that $f_{yy} = \alpha(x)y^c$ for $\alpha(x) > 0$ and $c \neq 0$. Equation (10.5.e) yields

$$\lambda = f_{yyy} f_{yy}^{-1} = cy^{-1}, \quad a_{12} = -f_{xyy} f_{yyy}^{-1} = -\tfrac{\alpha_x(x)y}{c\alpha(x)}, \quad \lambda^2 a_{11}^{-2} = f_{yy} = \alpha(x)y^c.$$

We apply Equation (10.5.f) to see

$$\begin{aligned} f_{xyyy} + a_{12} f_{yyyy} &= \alpha_x(x)cy^{c-1} - \tfrac{\alpha_x(x)y}{c\alpha(x)}c(c-1)\alpha(x)y^{c-2} = \alpha_x(x)y^{c-1} \\ &= a_{11}^{-3}\lambda^4 c_{122112} = \alpha(x)^{3/2}y^{3/2c}cy^{-1}c_{122112}. \end{aligned}$$

This implies that $\alpha_x(x)\alpha(x)^{-3/2} = c \cdot c_{122112} \cdot y^{c/2}$. Consequently, $\alpha_x(x) = 0$ so $\alpha(x) = a$ is constant. Therefore, $f_{yy} = ay^c$ for $c \neq 0$ and $a \neq 0$. Let $P(t)$ solve the equation $P''(t) = t^c$. We then have

$$f(x,y) = aP(y) + u(x)y + v(x).$$

We apply Equation (10.5.g) with $a_{12} = 0$:

$$\begin{aligned} a_{11}^{-4}\nabla^2 R(\xi_1, \xi_2, \xi_2, \xi_1; \xi_1, \xi_1) &= -f_y f_{yyy} = -aP'(y)acy^{c-1} - u(x)acy^{c-1} \\ &= c_{122111}\lambda^4 a_{11}^{-4} = c_{122111}a^2y^{2c}. \end{aligned}$$

If $c = -1$, then $P'(y) = \ln(y)$ and this relation is impossible. Consequently, $c \neq -1$ and we may conclude that $u(x) = 0$. We therefore have $f = aP(y) + v(x)$. As in Case I, the constant term is eliminated and a is set to 1 by making a change of variables

$$T(x, y, \tilde{x}) = (a^{-1/2}x, y, a^{1/2}\tilde{x} + 2w(x))$$

where $w_x(x) = v(x)$. Thus, $f = \pm\ln(y)$ or $f = \pm y^\varepsilon$ for $\varepsilon \neq 0, 1, 2$.

Case II-a. Let $f(y) = \ln(y)$; the case $f(y) = -\ln(y)$ is similar. We know by Theorem 10.25 that $\mathcal{M}_f$ is not 2-curvature homogeneous and, therefore, is not homogeneous. We clear the previous notation. For $\lambda > 0$ and $(x_0, \tilde{x}_0)$ arbitrary, set:

$$T(x, y, \tilde{x}) := (\lambda x + x_0, \lambda y, \lambda \tilde{x} + \tilde{x}_0 + (\lambda \ln \lambda) x)\,.$$

Let $\Theta := (x, y, \tilde{x})$ and $\tilde{\Theta} = T\Theta$. We compute:

$$\begin{aligned}
&T_*\partial_x = \lambda\partial_x + \lambda \ln \lambda \partial_{\tilde{x}}, \quad T_*\partial_y = \lambda\partial_y, \quad T_*\partial_{\tilde{x}} = \lambda\partial_{\tilde{x}},\\
&g(T_*\partial_x, T_*\partial_x)(\tilde{\Theta}) = \lambda^2\{-2\ln(y) - 2\ln\lambda\} + 2\lambda^2 \ln \lambda = \lambda^2 g(\partial_x, \partial_x)(P),\\
&g(T_*\partial_x, T_*\partial_y)(\tilde{\Theta}) = 0, \quad g(T_*\partial_x, T_*\partial_{\tilde{x}})(\tilde{\Theta}) = \lambda^2, \quad g(T_*\partial_y, T_*\partial_y)(\tilde{\Theta}) = \lambda^2,\\
&g(T_*\partial_y, T_*\partial_{\tilde{x}})(\tilde{\Theta}) = 0, \quad g(T_*\partial_{\tilde{x}}, T_*\partial_{\tilde{x}}) = 0\,.
\end{aligned}$$

This defines a transitive action by homotheties on $\mathbb{R} \times \mathbb{R}^+ \times \mathbb{R}$, which shows that $\mathcal{M}_{\ln(y)}$ is homothety homogeneous. Moreover, T acts by isometries on each level set of the projection $(x, y, \tilde{x}) \mapsto y$, showing that $\mathcal{M}_{\ln(y)}$ is of cohomogeneity one.

Case II-b. Let $f(y) = y^c$ for $c \neq 0, 1, 2$. Again, Theorem 10.25 implies that $\mathcal{M}_f$ is not 2-curvature homogeneous and, therefore, not homogeneous. Let

$$T(x, y, \tilde{x}) = (\lambda^{(2-c)/2}x + x_0, \lambda y, \mp\lambda^{2+(c-2)/2}\tilde{x} + \tilde{x}_0)\,.$$

We compute:

$$\begin{aligned}
&T_*\partial_x = \lambda^{(2-c)/2}\partial_x, \quad T_*\partial_y = \lambda\partial_y, \quad T_*\partial_{\tilde{x}} = \lambda^{2+(c-2)/2}\partial_{\tilde{x}},\\
&g(T_*\partial_x, T_*\partial_x)(T(x, y, \tilde{x})) = -2\lambda^{(2-c)}\lambda^c y^c = \lambda^2 g(\partial_x, \partial_x)(x, y, \tilde{x}),\\
&g(T_*\partial_x, T_*\partial_y) = 0, \quad g(T_*\partial_x, T_*\partial_{\tilde{x}}) = \lambda^2, \quad g(T_*\partial_y, T_*\partial_y) = \lambda^2,\\
&g(T_*\partial_y, T_*\partial_{\tilde{x}}) = 0, \quad g(T_*\partial_{\tilde{x}}, T_*\partial_{\tilde{x}}) = 0\,.
\end{aligned}$$

Thus, T is a homothety; since $\lambda > 0$ is arbitrary and since $(x_0, \tilde{x}_0)$ are arbitrary, the group of homotheties acts transitively on M. This completes the proof of Theorem 10.27. □

10.6 STABILITY

This section is devoted to the proof of the following result.

Theorem 10.29 *Let $\mathcal{M} = (M, g)$ be a pseudo-Riemannian manifold. The following assertions are equivalent.*

(a) $\mathcal{M}$ is locally homogeneous.

(b) $\mathcal{M}$ is k-curvature homogeneous for all k.

(c) $\mathcal{M}$ is k-curvature homogeneous for $k = \frac{1}{2}m(m-1)$.

The condition that $\mathcal{M}$ is k-curvature homogeneous for all k is often referred to as *infinitesimally homogeneous* in the literature.

The proof of Theorem 10.29 will occupy the remainder of this section. We follow the original work of Podestà and Spiro [120] which generalized a result of Singer [125] to a more general context of G-structures (see also work of Pecastaing [114] for further generalizations). In Section 10.6.1, we introduce the orthogonal group and define the *Singer number*. In Section 10.6.2 we review some facts concerning principal bundles. In Section 10.6.3, we give conditions under which morphisms of principal bundles arise from corresponding morphisms of the base. Section 10.6.4 treats the equivalence problem where the structure group is the trivial group. This turns out, somewhat surprisingly, to be the crucial case. In Section 10.6.5 we complete the proof of Theorem 10.29.

10.6.1 THE SINGER NUMBER.

Let $\{e_1, \dots, e_m\}$ be the standard basis for $\mathbb{R}^m$ and let $\langle\cdot,\cdot\rangle_{p,q}$, for $p+q=m$, be the standard inner product of signature (p,q). Let

$$O(V, \langle\cdot\,,\,\cdot\rangle) := \{T \in \mathrm{GL}(m, \mathbb{R}^m) : T^*\langle\cdot,\cdot\rangle_{p,q} = \langle\cdot,\cdot\rangle_{p,q}\}\,. \tag{10.6.a}$$

This is the orthogonal group of dimension $\frac{1}{2}m(m-1)$. Let $\mathfrak{o}(p,q)$ be the associated Lie algebra. We will also use the notation $\mathcal{O}(p,q)$, to emphasize the signature (p,q) of the inner product and let $\mathfrak{o}(p,q)$ be the associated Lie algebra. To simplify the notation, et $\mathcal{O}(m) = \mathcal{O}(0,m)$.

Definition 10.30 If P is a point of a pseudo-Riemannian manifold $\mathcal{M} = (M, g)$, let

$$\mathcal{O}_P^s(\mathcal{M}) := \left\{T \in O(T_PM, g_P) : T^*(\nabla^\ell R_P) = \nabla^\ell R_P \quad \text{for} \quad 0 \le \ell \le s\right\}. \tag{10.6.b}$$

This is the Lie group of isometries of (T_PM, g_P) which fix the s-first covariant derivatives of the curvature operator. Let $a \cdot \nabla^\ell\mathfrak{R}$ be the extension of a to act as a derivation; $a \cdot \nabla^\ell\mathfrak{R}(x_1, x_2, \dots) = -\nabla^\ell\mathfrak{R}(ax_1, x_2, \dots) - \nabla^\ell\mathfrak{R}(x_1, ax_2, \dots) - \cdots$. Let $\mathfrak{o}_P^s(\mathcal{M})$ be the associated Lie algebra;

$$\mathfrak{o}_P^s(\mathcal{M}) := \left\{a \in \mathfrak{o}(T_PM, g_P) : a \cdot \nabla^\ell\mathfrak{R} = 0 \text{ for } 0 \le \ell \le s\right\}, \tag{10.6.c}$$

Clearly, one has inclusions $\mathfrak{o}_P^{s+1}(\mathcal{M}) \subset \mathfrak{o}_P^s(\mathcal{M})$. Choose $s_{\mathcal{M}}^O(P)$ minimal so

$$\mathfrak{o}_P^s(\mathcal{M}) = \mathfrak{o}_P^{s+1}(\mathcal{M}) \quad \text{for} \quad s \ge s_{\mathcal{M}}^O(P)\,.$$

By Lemma 9.21, if $\mathcal{M}$ is k-curvature homogeneous, then $\mathfrak{o}_P^\ell(\mathcal{M})$ and $\mathfrak{o}_Q^\ell(\mathcal{M})$ are conjugate for all $\ell \le k$ and for any $P, Q \in M$. Hence, if the structure $(\mathcal{M}, g)$ is k-curvature homogeneous for $k = s_{\mathcal{M}}(P) + 1$, then $s_{\mathcal{M}}(P)$ does not depend on $P \in M$. In this setting, we let $s_{\mathcal{M}} := s_{\mathcal{M}}(P)$ for any P and refer to $s_{\mathcal{M}}^{\mathcal{HO}}$ as the *Singer number*. Since

$$\dim(\mathcal{O}(V, \langle\cdot,\cdot\rangle)) = \tfrac{1}{2}m(m-1) \quad \text{we have} \quad s_{\mathcal{M}} \le \tfrac{1}{2}m(m-1) - 1\,.$$

10.6.2 PRINCIPAL BUNDLES. We introduced the notion of a *fiber bundle* in Section 2.1.9 of Book I and we refer to the discussion there for further details.

Definition 10.31 A *principal G bundle* is a fiber bundle $\pi : \mathcal{P} \to M$ where the fiber is a Lie group G. We assume in addition given a smooth proper fiber preserving right fixed point free action of G on $\mathcal{P}$ so that if ξ is section to $\mathcal{P}$ over an open subset $\mathcal{O}$ of M, then the map $(x, \theta) \to \xi(x)\theta$ is a fiber preserving diffeomorphism from $\mathcal{O} \times G$ to $\pi^{-1}(\mathcal{O})$ which gives local trivializations for $\mathcal{P}$. We will also denote this action by R_θ where convenient. The transition functions between two such local trivializations $\xi_1 = \xi_2\theta$ are then given by left multiplication by θ on the fiber G since identifying $\xi_1\theta_1 = \xi_2\theta_2$ means identifying $\theta\theta_1 = \theta_2$. On the other hand, given an open cover $\mathcal{O}_\alpha$ of M and smooth functions $\theta_{\alpha\beta} : \mathcal{O}_\alpha \cap \mathcal{O}_\beta \to G$ satisfying the *cocycle condition* $\theta_{\alpha\beta}\theta_{\beta\gamma} = \theta_{\alpha\gamma}$ on $\mathcal{O}_\alpha \cap \mathcal{O}_\beta \cap \mathcal{O}_\gamma$ (see Equation (2.1.a) of Book I), we can construct a principal bundle by identifying (ξ, θ_α) over $\mathcal{O}_\alpha$ with (ξ, θ_β) over $\mathcal{O}_\beta$ if $\theta_\alpha = \theta_{\alpha\beta}\theta_\beta$.

Example 10.32 Let $\mathrm{GL}(k, \mathbb{R})$ be the Lie group of linear transformations of $\mathbb{R}^k$. If W is a k-dimensional vector real bundle over M, let $\mathcal{F}(W)$ be the bundle of frames for W. The fiber of $\mathcal{F}(W)$ over a point P of M is the set of bases $\xi = \{e_1, \dots, e_k\}$ for W_P. This is a principal $\mathrm{GL}(k, \mathbb{R})$ bundle where $\xi\theta = (e_i\theta_1^i, \dots, e_i\theta_m^i)$ for $\theta_j^i \in \mathrm{GL}(k, \mathbb{R})$. If W is a complex vector bundle, then the structure group is $\mathrm{GL}(k, \mathbb{C})$.

Definition 10.33 Let $\mathcal{P}$ be a principal G bundle over M. Given a left linear action of G on a vector space V, we can define the vector bundle $\mathcal{P} \times_G V$ over P to be the Cartesian product $\mathcal{P} \times V$ by identifying $(\xi\theta, v) = (\xi, \theta v)$ and setting $\pi(\xi, v) := \pi\xi$.

If W is a vector bundle over M, then the vector bundle associated with the usual action of $\mathrm{GL}(k, \mathbb{R})$ on $\mathbb{R}^k$ is canonically isomorphic to W, the vector bundle associated to the natural dual action of $\mathrm{GL}(k, \mathbb{R})$ on $(\mathbb{R}^k)^*$ is W^*, and so forth. Thus, the geometry of all these bundles is encoded in the frame bundle.

Definition 10.34 A *principal bundle morphism* is a bundle morphism $f : \mathcal{P}_1 \to \mathcal{P}_2$ so that $f(\xi\theta) = f(\xi)\phi(\theta)$ where ϕ is a Lie group homomorphism from the structure group G_1 of $\mathcal{P}_1$ to the structure G_2 of $\mathcal{P}_2$. If G_1 is a Lie subgroup of G_2 and if ϕ and f are inclusions, then we say that f is an *embedding of principal bundles* and that we have a *reduction of the structure group* of $\mathcal{P}_2$ from G_2 to G_1. A *G-structure* on M is a reduction of the structure group of the tangent bundle of M to a subgroup G of $\mathrm{GL}(m, \mathbb{R})$.

Let (M, g) be a pseudo-Riemannian manifold of dimension m. Let $\mathcal{F}_{O(p,q)}(TM)$ be bundle of orthonormal frames. The fiber of $\mathcal{F}_{O(p,q)}(TM)$ over a point P of M is the set of orthonormal frames $\xi = \{e_1, \dots, e_m\}$ for $T_P M$ with the action of Example 10.32 where $g(e_i, e_j) = 0$ for

$i \neq j, g(e_i, e_i) = -1$ for $i \leq p$, and $g(e_i, e_i) = +1$ for $i > p$. This subbundle of the frame bundle of TM gives a reduction of the structure group from the general linear group $\mathrm{GL}(m, \mathbb{R})$ to the orthogonal group $O(p, q)$ and defines an $O(p, q)$-structure on M. Suppose that (M, g, J) is a complex manifold which is equipped with a J-invariant Riemannian metric g of dimension $2\bar{m}$. The $U(\bar{m})$-frame bundle is a $U(\bar{m})$ principal bundle and the natural inclusion of this bundle into $\mathcal{F}_{O(m)}(TM)$ is an embedding and defines a reduction of the structure group from $O(m)$ to $U(\bar{m})$ and defines a $U(\bar{m})$-structure on M.

Definition 10.35 We say that M is *parallelizable* if the frame bundle $\mathcal{F}(TM)$ is trivial, i.e., if there exist globally defined vector fields $\{e_1, \ldots, e_m\}$ spanning the tangent space at every point. Every Lie group is parallelizable. More generally, a vector bundle W is trivial if and only if $\mathcal{F}(W)$ admits a global section or, equivalently, $\mathcal{F}(W)$ admits a reduction to the trivial structure group $\{\mathrm{Id}\}$.

A manifold M is parallelizable if and only if it is possible to reduce the structure group of the frame bundle to the trivial group $\{\mathrm{Id}\}$. Not every manifold is globally parallelizable, although this condition can always be satisfied in any coordinate chart since the coordinate vector fields give a framing. Let $A \to 1 \oplus A$ embed $\mathrm{GL}(m-1, \mathbb{R})$ into $\mathrm{GL}(m, \mathbb{R})$. Then M admits a reduction of the structure group to $\mathrm{GL}(m-1, \mathbb{R})$ if and only if M admits a nowhere vanishing vector field. If M is compact and without boundary, this is equivalent to the vanishing of the *Euler–Poincaré characteristic* $\chi(M)$.

Definition 10.36 Let $\mathcal{P}$ be a principal bundle. The *vertical space* $\mathfrak{V} := \ker\{\pi_*\}$ is a canonically defined subbundle of the tangent bundle of $\mathcal{P}$. Furthermore, the G action identifies the fiber of $\mathfrak{V}$ with the Lie algebra $\mathfrak{g}$ of G. There is no canonically defined complementary bundle. A *connection on a principal bundle* $\mathcal{P}$ is a G-invariant splitting $T(\mathcal{P}) = \mathfrak{V} \oplus \mathfrak{H}$. A principal bundle connection can be described in terms of the associated *connection form*. Define a $\mathfrak{g}$-valued 1-form ω on $\mathcal{P}$ by setting $\omega(v) = v$ if $v \in \mathfrak{V}$ and $\omega(h) = 0$ if $h \in \mathfrak{H}$. Let $R_\theta : \xi \to \xi\theta$ denote the action of G on $\mathcal{P}$. The fact that the splitting is G-invariant means $R_\theta^* \omega = \mathrm{ad}(\theta^{-1})\omega$. On the other hand, any 1-form with these properties defines an equivariant splitting if we take $\mathfrak{H} = \ker\{\omega\}$. We may therefore define a connection to be a $\mathfrak{g}$-valued 1-form on $\mathcal{P}(M)$ so that $\omega(v \oplus h) = v \quad$ and $\quad R_g^* \omega = ad(g^{-1})\omega$ for any $A \in \mathfrak{G} \quad$ and $\quad \theta \in G$.

Example 10.37 Let ∇ be a connection on a vector bundle W over M. Let $\vec{x} = (x^1, \ldots, x^m)$ be local coordinates on M and let $\vec{s} = (s^1, \ldots, s^k)$ be a local frame for W. We obtain local coordinates $(\vec{x}, \theta)$ for $\theta = \theta_a^b \in \mathrm{GL}(k, \mathbb{R}) \subset \mathbb{R}^{k^2}$ on the frame bundle $\mathcal{F}(W)$. We obtain an invariantly defined connection on $\mathcal{F}(TM)$ by setting $\mathfrak{H} = \mathrm{span}\{\partial_{x^i} - \Gamma_{ia}{}^b \partial_{\theta_a^b}\}$. Conversely, given a connection on $\mathcal{F}(W)$, one can recover a linear connection on W using this formalism. By using the Levi–Civita connection of a pseudo-Riemannian manifold (M, g), we obtain a connection on $\mathcal{F}_{O(p,q)}(TM)$.

Definition 10.38 Let $\xi = (e_1, \dots, e_m) \in \mathcal{F}(TM)$ and let $\pi : \mathcal{F}(TM) \to M$ be the canonical projection. If $X \in T_\xi \mathcal{P}(M)$, expand $\pi_*(X) = a^i e_i$ to define $\vartheta(X) = (a^1, \dots, a^m) \in \mathbb{R}^m$. This gives a canonically defined 1-form ϑ on $\mathcal{P}$ with values in $\mathbb{R}^m$ which satisfies the relation (see, for example, Sternberg [128])

$$R_\theta^* \vartheta = \theta^{-1} \vartheta \quad \text{and} \quad d\vartheta(A^*, \cdot) = -A\,\vartheta.$$

10.6.3 PROJECTING PRINCIPAL BUNDLE MORPHISMS TO THE BASE. Any diffeomorphism $f : M_1 \to M_2$ lifts to a bundle isomorphism $\hat{f}$ of the associated tangent frame bundles if we set $\hat{f}(P, e_1, \dots, e_m) := (f(P), f_*(P)e_1, \dots, f_*(P)e_m)$. The following result (see Kobayashi and Nomizu [85, Chapter VI]) gives a criteria in terms of the canonical 1-form which ensures that a bundle isomorphism of the frame bundles arises from a diffeomorphism of the base:

Lemma 10.39 *Let $\mathcal{F}_i = \mathcal{F}(TM_i)$ be the principal frame bundles of m-dimensional manifolds M_i for $i = 1, 2$.*

1. *Let f be a diffeomorphism from M_1 to M_2 and let $\hat{f}$ be the induced principal bundle morphism from $\mathcal{F}(TM_1)$ to $\mathcal{F}(TM_2)$. Then $\hat{f}^* \vartheta_2 = \vartheta_1$.*
2. *Let $\hat{f} : \mathcal{F}(TM_1) \to \mathcal{F}(TM_2)$ be a fiber preserving principal bundle morphism which satisfies $\hat{f}^* \vartheta_2 = \vartheta_1$. Then $\hat{f}$ is induced by an underlying diffeomorphism f from M_1 to M_2.*

Let M_i be equipped with G-structures $\mathcal{P}_{i,G}$. We say these two G-structures are *equivalent G-structures* if there is a diffeomorphism $f : M_1 \to M_2$ so $\hat{f} : \mathcal{P}_{1,G} \to \mathcal{P}_{2,G}$ is a principal bundle diffeomorphism. We can restrict the canonical 1-form ϑ to $\mathcal{P}_{i,G}$. Clearly, $\hat{f}^* \vartheta_2 = \vartheta_1$ in this setting. Furthermore, in light of Lemma 10.39, a principal bundle diffeomorphism $\hat{f} : \mathcal{P}_{1,G} \to \mathcal{P}_{2,G}$ arises from an equivalence of G-structures if and only if $\hat{f}$ is G-equivariant and preserves ϑ. If $G = O(p,q)$, this means that f is an isometry, if $G = U(\bar{m})$, this means that f is a unitary map, and so forth. Thus, this reduces the problem of constructing isometries to the problem of constructing principle bundle diffeomorphisms which preserve ϑ.

Definition 10.40 If f is a principal bundle map from $\mathcal{P}(M)$ to $\mathcal{P}(M)$, then $f^* \mathfrak{H}$ is again a connection on $\mathcal{P}$. Let $\mathfrak{A}$ be a family of such automorphisms. We say that a connection is a *canonical connection* if it is invariant under elements of $\mathfrak{A}$; we omit $\mathfrak{A}$ from the notation in the interests of notational simplicity.

Definition 10.41 We now pass one level higher. Let G be a subgroup of the general linear group and let $\mathcal{P}$ be a reduction of the structure group of the tangent bundle to G. Assume $\mathcal{P}$ is equipped with a connection $\mathfrak{H}$ with associated connection 1-form ω. Then the one-form $\omega \oplus \vartheta$, at each point $\xi \in \mathcal{P}_G(M)$ defines an isomorphism $(\omega \oplus \vartheta)_\xi : T_\xi \mathcal{P}_G(M) \to \mathfrak{g} \oplus \mathbb{R}^m$. Consequently, by considering $\mathfrak{g} \oplus \mathbb{R}^m$ as the model vector space of $T_\xi \mathcal{P}_G(M)$, $(\omega \oplus \vartheta)_\xi^{-1}$ is a frame at $\xi \in \mathcal{P}_G(M)$; we see that $\mathfrak{H}$ determines a section $\sigma_{\mathfrak{H}} : \mathcal{P}_G(M) \to \mathcal{F}(\mathcal{P}_G(M))$. This equivalence between connections and global sections relates the automorphisms of a G-structure with a canonical connection (with respect to some family of automorphisms) and diffeomorphisms which respect the corresponding global section.

10.6.4 THE EQUIVALENCE PROBLEM FOR {Id}-STRUCTURES. The case in which $G = \{\mathrm{Id}\}$ plays a distinguished role in understanding the automorphims of many G-structures. Cartan introduced the notion of the *prolongation* of a G-structure to study the group of automorphisms of a principal G bundle. It is a canonical procedure which associates to each subgroup G of the general linear group an abelian subgroup $G^{(1)} \subset GL(\mathfrak{g} \oplus \mathbb{R}^k)$. There is, correspondingly, a $G^{(1)}$-structure on the principal bundle so that an automorphism ϕ of $\mathcal{P}$ induces an automorphism $\hat{\phi} \in \mathrm{Aut}(\mathcal{P}_G(M), \mathcal{P}_{G^{(1)}}(\mathcal{P}_G(M)))$ if and only if $\phi = \hat{f}$ for some automorphism f. Iterating this construction get a sequence of groups $G^{(2)}, \ldots$ with, in some special cases, $G^{(k)} = \{\mathrm{Id}\}$. See, for example, Sternberg [128]. We wish to study the orbits of the automorphisms of an {Id}-structure.

Definition 10.42 Let $\mathfrak{F} \subset C^\infty(M)$ be a collection of functions on a manifold M. Set $\mathrm{Rank}(\mathfrak{F}_P) := \dim(\mathrm{span}\{df_P\}_{f \in \mathfrak{F}})$. $\mathfrak{F}$ is said to be *regular at* P if $\mathrm{Rank}(\mathfrak{F}_Q)$ is constant near P. If this condition holds, let $k = \mathrm{Rank}_P(\mathfrak{F})$. Choose functions $f^i \in \mathfrak{F}$ for $1 \le i \le k$ so $\{df^1(P), \ldots, df^k(P)\}$ are linearly independent. We may then choose local coordinates $(x^1, \ldots, x^m)$ near P so that $x^i = f^i$ for $1 \le i \le k$. Such a coordinate system is said to be an *adapted coordinate system* for $\mathfrak{F}$.

The following observation is now immediate.

Lemma 10.43 Let $\vec{x}$ be an adapted coordinate system for a regular family $\mathfrak{F}$ of functions with $\mathrm{Rank}(\mathfrak{F}_P) = k$. If $f \in \mathfrak{F}$, then df is a linear combination of $\{dx^1, \ldots, dx^k\}$ and, consequently, f has the form $f = f(x^1, \ldots, x^k)$.

Definition 10.44 Suppose that $\sigma = (X_1, \ldots, X_m)$ is a global section to the frame bundle $\mathcal{F}(TM)$. Let $\{\omega^1, \ldots, \omega^m\}$ be the dual frame for T^*M. The *component functions* $c_{ij}{}^k$ of the *torsion*

$$c_\sigma : M \to \mathrm{Hom}(\Lambda^2(\mathbb{R}^m), \mathbb{R}^m)$$

are characterized by the identity

$$c_\sigma(e_i \wedge e_j) = c_{ij}{}^k e_k = \omega^k([X_i, X_j])e_k = -d\omega^k(X_i, X_j)e_k.$$

The functions $c_{ij}{}^k$ are called the (first-order) *structure functions* of σ. We denote this family of $\frac{1}{2}m^2(m-1)$ functions by $\mathfrak{F}^0$. Let $\mathfrak{F}^1$ be the family of

$$\frac{1}{2}m(m+m^2)(m-1)$$

functions given by $\mathfrak{F}^1 = \{c_{jk}{}^i, c_{jk;\ell}{}^i\}$. Let ";" denote covariant differentiation. More generally, we define:

$$\mathfrak{F}^s = \{c_{jk}{}^i,\ c_{jk;\ell_1}{}^i,\ \dots, c_{jk;\ell_1\dots\ell_s}{}^i\}.$$

Since $\mathfrak{F}^s \subset \mathfrak{F}^{s+1}$,

$$0 \le \operatorname{Rank}(\mathfrak{F}^0_P) \le \operatorname{Rank}(\mathfrak{F}^1_P) \le \dots \le \operatorname{Rank}(\mathfrak{F}^s_P) \le m\,.$$

We refer to Sternberg [128] for the proof of the following stability result.

Lemma 10.45 *Suppose that $\mathfrak{F}^{s+1}$ is regular at $P \in M$. If $\operatorname{Rank}(\mathfrak{F}^s_P) = \operatorname{Rank}(\mathfrak{F}^{s+1}_P)$, then $\operatorname{Rank}(\mathfrak{F}^s_P) = \operatorname{Rank}(\mathfrak{F}^{s+\nu}_P)$ for all ν.*

Definition 10.46 A frame $\sigma = (X_1, \dots, X_m)$ for TM is said to be a *regular frame* at $P \in M$ if there exists s so that $\mathfrak{F}^{s+1}$ is regular at $P \in M$ and if $\operatorname{Rank}(\mathfrak{F}^s_P) = \operatorname{Rank}(\mathfrak{F}^{s+1}_P)$. The smallest such s is called *the order of the parallelism at $P \in M$* and the integer $\operatorname{Rank}(\mathfrak{F}^s_P)$ is called *the rank of the parallelism at $P \in M$*.

It follows from Theorem 4.1 in Sternberg [128, Chapter VII] that the rank and the order determine the local equivalence problem for regular complete parallelisms. The following result will turn to be essential in the proof of Theorem 10.29.

Theorem 10.47 *Let M be a manifold with a complete parallelism which is regular at some $P \in M$ with constant rank k in a sufficiently small neighborhood of P. Let $(x^1, \dots, x^m)$ be a coordinate system adapted to $\mathfrak{F}^{k+1}$. Then, in a neighborhood of P, any point on the manifold*

$$x^1 = \text{const}, \dots, x^k = \text{const}\,,$$

can be carried out into any other such point by a local automorphism of M.

10.6.5 THE PROOF OF THEOREM 10.29. The Levi–Civita connection defines a natural connection on the principal orthogonal frame bundle $\mathcal{P}_{O(p,q)}$ which has connection 1-form ω (see Example 10.37). Let ϑ be the canonical 1-form defined in Definition 10.38. At $\xi \in \mathcal{P}_{O(p,q)}$, we use the discussion in Definition 10.41 to see that $\omega \oplus \vartheta$ defines an isomorphism $(\omega \oplus \vartheta)_\xi : T_\xi \mathcal{P}_G(M) \to \mathfrak{o}(p,q) \oplus \mathbb{R}^m$. Let $\mathfrak{o}(p,q) \oplus \mathbb{R}^m$ be the model vector space of $T_\xi \mathcal{P}_{O(p,q)}(M)$. Then $(\omega \oplus \vartheta)_\xi^{-1}$ is a frame at $\xi \in \mathcal{P}_{O(p,q)}(M)$. Therefore, the connection $\mathfrak{H}$ determines a section

$$\sigma_{\mathfrak{H}} : \mathcal{P}_{O(p,q)}(M) \to \mathcal{F}(\mathcal{P}_{O(p,q)}(M)) .$$

Let $\{e_1, \ldots, e_m\}$ be the standard basis for $\mathbb{R}^m$. Consider on $\mathbb{R}^m$ the standard inner product of signature (p,q), for $p+q=m$. Let $\{e^1, \ldots, e^m\}$ be the dual basis. Let

$$\epsilon_{ij} = \epsilon^{ij} = g(e_i, e_j) .$$

Let $E_i^j = e_i \otimes e^j$ be the canonical basis of the Lie algebra $\mathfrak{o}(p,q)$. Then $\omega = \omega_j^i\, E_i^j$ is the $\mathfrak{o}(p,q)$-valued one-form on $\mathcal{P}_{O(p,q)}(M)$ defined by the Levi–Civita connection of $\mathcal{M}$. Let $E_i^{j\,*}$ be the fundamental vector field on $\mathcal{P}_{O(p,q)}(M)$ corresponding to $E_i^j \in \mathfrak{o}(V, \langle \cdot, \cdot \rangle)$ (see Example 10.37). One then has $\omega_r^s(E_i^{j\,*}) = \delta_{rj}\delta_{is}$.

We work locally in a coordinate chart $\mathcal{U}$. Let $\{X_1, \ldots, X_m\}$ be a local orthonormal frame over $\mathcal{U}$ whose lift $\{\mathbb{X}_1, \ldots, \mathbb{X}_m\}$ to $\mathcal{P}_{O(p,q)}(\mathcal{U})$ with respect to the Levi–Civita connection satisfies $\vartheta^j(\mathbb{X}_i) = \delta_{ij}$. One then has

$$\vartheta^k(\mathbb{X}_i) = \delta_{ik}\,, \quad \vartheta^k(E_i^{j\,*}) = 0\,, \quad \omega_\ell^k(\mathbb{X}_i) = 0\,, \quad \omega_\ell^k(E_i^{j\,*}) = \delta_{ik}\delta_{\ell j}\,.$$

Hence, $\{\vartheta^i, \omega_j^i\}$ is a complete parallelism on $\mathcal{P}_{O(p,q)}(\mathcal{U})$. Let $R_{k\ell j}{}^i$ be the curvature tensor of the Levi–Civita connection. Since the Levi–Civita connection is torsion-free, the structure equations (see Kobayashi and Nomizu [85]) take the form

$$d\vartheta^i = -\omega_j^i \wedge \vartheta^j\,, \quad d\omega_j^i = -\omega_k^i \wedge \omega_j^k + \tfrac{1}{2}\mathfrak{R}_{k\ell j}^i \vartheta^k \wedge \vartheta^\ell\,.$$

Hence, the structure functions of the parallelism of $\mathcal{P}_{O(p,q)}(\mathcal{U})$ determined by the Levi–Civita connection are given by

$$\vartheta^k([\mathbb{X}_i, \mathbb{X}_j]) = -d\vartheta^k(\mathbb{X}_i, \mathbb{X}_j) = 0,$$

$$\vartheta^k([\mathbb{X}_i, E_j^{\ell *}]) = -d\vartheta^k(\mathbb{X}_i, E_j^{\ell *}) = \delta_{i\ell}\delta_{jk},$$

$$\vartheta^k([E_i^{r*}, E_j^{\ell *}]) = -d\vartheta^k(E_i^{r*}, E_j^{\ell *}) = 0,$$

$$\omega_\ell^k([\mathbb{X}_i, \mathbb{X}_j]) = -d\omega_\ell^k(\mathbb{X}_i, \mathbb{X}_j) = R_{ji\ell}{}^k, \qquad \omega_\ell^k([\mathbb{X}_i, E_j^{\ell *}]) = 0,$$

$$\omega_\ell^k([E_i^{r*}, E_j^{\nu *}]) = -d\omega_\ell^k(E_i^{r*}, E_j^{\nu *}) = \omega_s^k \wedge \omega_\ell^s(E_i^{r*}, E_j^{\nu *}) = \delta_{ik}\delta_{jr}\delta_{\ell\nu} - \delta_{jk}\delta_{i\nu}\delta_{\ell r}\,.$$

Let $\mathbb{X}_s(R_{ij\ell}{}^k)$ be the derivative of the function $R_{ij\ell}{}^k$ on $\mathcal{P}_{O(p,q)}(\mathcal{U})$ with respect to the lifted vector field $\mathbb{X}_s$ and let $(\nabla_{\mathbb{X}_s}\mathfrak{R})_{ij\ell}{}^k$ be the covariant derivative on the base (M,g). Only the structure equations $\omega_\ell^k([\mathbb{X}_i,\mathbb{X}_j]) = R_{ji\ell}{}^k$ are non-constant. Thus, all the derivatives of the structure equations vanish except for the following

$$\mathbb{X}_s(R_{ij\ell}{}^k) = (\nabla_{\mathbb{X}_s}\mathfrak{R})_{ij\ell}^k \quad \text{and} \quad E_s^{r*}(R_{ij\ell}{}^k) = E_s^r \cdot R_{ij\ell}{}^k . \tag{10.6.d}$$

Hence, the higher-order derivatives of the structure functions are completely determined by the corresponding covariant derivatives of the curvature tensor and the action of the E_s^r's on the lower order covariant derivatives of the curvature.

Let $\mathcal{M}$ be infinitesimally homogeneous. Let P and Q be points of $\mathcal{U}$. Let $s_{\mathcal{M}}$ be the Singer number discussed in Definition 10.30. Choose a linear isometry $\Phi_{P,Q}$ from $T_P M$ to $T_Q M$ so that $\Phi_{P,Q}^*(\nabla^\ell\mathfrak{R}_Q) = \nabla^\ell\mathfrak{R}_P$ for $\ell \le s_{\mathcal{M}}$. We will show that $\Phi_{P,Q}$ arises from the germ of an isometry $\phi_{P,Q}$ of the underlying manifold (M,g); the equivalences of Theorem 10.29 will then follow.

Let $\mathcal{U}$ be a small coordinate neighborhood. We introduce some additional notation. Fix a non-negative integer α. Let $\mathfrak{F}^\alpha$ be the family of all the structure functions and their derivatives up to order α. Let $\nabla^\alpha R_P$ be the covariant derivatives of the curvature tensor up to order α at $P \in M$. Adopt the notation of Equation (10.6.b) and Equation (10.6.c) to define $\mathcal{O}_P^s(\mathcal{M})$ and $\mathfrak{o}_P^s(\mathcal{M})$. Let $\xi \in \mathcal{P}_{O(p,q)}(\mathcal{U})$ with $\pi(\xi) = P$. Then iterating the procedure of Equation (10.6.d) yields

$$\operatorname{Rank}(\mathfrak{F}_\xi^\alpha) = \dim(\mathfrak{o}(T_P M, g_P)) - \dim(\mathfrak{o}^{\alpha+1}(T_P M, g_P)) + \operatorname{Rank}(\nabla^{\alpha+1} R_P) .$$

Consequently, if (M,g) is infinitesimally homogeneous, then each ortonormal frame $\xi \in \mathcal{P}_{O(p,q)}(\mathcal{U})$ is a regular point for the parallelism. Moreover, its order is exactly the Singer number $s_{\mathcal{M}}^{\mathcal{O}}$ (see Definition 10.30). Hence, around any point $\xi \in \mathcal{P}_{O(p,q)}(\mathcal{U})$ there exist adapted coordinates (x^i) such that the first r-coordinates are given by r-linearly independent structure functions in $\mathfrak{F}^{s_{\mathcal{M}}}$ (where $r = \operatorname{Rank}(\mathfrak{F}^{s_{\mathcal{M}}})$). Theorem 10.47 now shows that, in such a coordinate neighborhood, any point $\bar{\xi} \in \mathcal{P}_{O(p,q)}(\mathcal{U})$ lying on the manifold $x^1 = \text{const}, \dots, x^r = \text{const}$ can be carried out into ξ by a local automorphism of $\mathcal{P}_{O(p,q)}(\mathcal{U})$ preserving the parallelism.

Hence, for each point $P \in \mathcal{U}$ there exists $\xi \in \mathcal{P}_{O(p,q)}(\mathcal{U})$ with $\pi(\xi) = P$ and an open neighborhood $\xi \in \mathcal{V} \subset \mathcal{P}_{O(p,q)}(\mathcal{U})$ which admits an adapted coordinate system $(x^1, \dots, x^N)$. Set $c_1 = x^1(\xi), \dots, c_r = x^r(\xi)$. Observe that since the first r-adapted coordinates are given by some linearly independent functions of the family $\mathfrak{F}^{s_{\mathcal{M}}}$, they are completely determined by the curvature and its covariant derivatives up to order $\ell \le s_{\mathcal{M}}^{\mathcal{O}} + 1$. Now, for each point $Q \in \pi(\mathcal{V})$ there exists an orthonormal frame $\bar{\xi} \in \mathcal{P}_{O(p,q)}(\mathcal{U})$, $\pi(\bar{\xi}) = Q$ with the same curvature components as ξ since, by assumption, (M,g) is infinitesimally homogeneous. Consequently, $x^1(\bar{\xi}) = c_1, \dots, x^r(\bar{\xi}) = c_r$. This shows that there exists an automorphism ϕ of $\mathcal{P}_{O(p,q)}(\mathcal{U})$ with $\phi(\xi) = \bar{\xi}$ such that it preserves the parallelism given by $\{\vartheta^i, \omega_j^i\}$. Hence, ϕ preserves the canonical form ϑ of $\mathcal{P}_{O(p,q)}(\mathcal{U})$ (i.e., $\phi^*\vartheta = \vartheta$) and, moreover, since it preserves the one-forms ω_j^i, it preserves the connection one-form ω. Now, since $\phi^*\omega = \omega$, ϕ maps fibers into fibers. Then Lemma 10.39 (see also the discussion in Definition 10.35) shows that ϕ is the lift of a diffeomorphim $f : \mathcal{U} \to \mathcal{U}$ with $f(P) = Q$. Any such diffeomorphism is necessarily an isometry. Since M is connected, a finite number of iterations shows that $\mathcal{M}$ is locally homogeneous. □

10.7 LOCALLY HOMOGENEOUS METRIC G-STRUCTURES

In Section 9.2, we considered invariants where the structure group was the orthogonal group. In this section, we follow the discussion of Díaz-Ramos, García-Río and Nicolodi [56] to consider more general invariants. Let $\mathcal{M} := (M, g)$ be a Riemannian manifold of dimension n. If $\mathcal{M}$ is locally homogeneous, then all the local scalar invariants (see the discussion in Section 9.2.1) are necessarily constant. Prüfer, Tricerri and Vanhecke [123] showed, conversely, that if $\mathcal{M}$ is a Riemannian manifold of dimension m and if all the local scalar invariants up to order $\frac{m(m-1)}{2} + 1$ are constant, then $\mathcal{M}$ is locally homogeneous and the local isometry type of M is uniquely determined by these curvature invariants. Again, note that a universal bound was given in terms of the dimension. In Theorem 10.29, we gave similar universal bounds involving curvature homogeneity. Other generalizations of this result have been obtained by Console and Nicolodi [51] and Opozda [110], among others.

Imposing additional structures on $\mathcal{M}$ reduces the structure group. For example, if $\mathcal{M}$ is an almost Hermitian manifold of complex dimension $\bar{n}$, then the natural structure group is the unitary group $U(\bar{n})$ rather than the orthogonal group $O(n = 2\bar{n})$. More generally, fix a closed subgroup $G \subset O(n)$ and assume the structure group can be reduced from $O(n)$ to G; this gives rise to the notion of a G-structure on $\mathcal{M}$. Reducing the structure group creates additional invariants. This section is devoted to the proof of the following result. It is a Riemannian result as the existence of VSI manifolds and the need to construct invariants which are not of Weyl type presented in Section 9.5 shows that no such theorem can apply in the pseudo-Riemannian setting.

Theorem 10.48 *A Riemannian manifold M is locally G-homogeneous if and only if all the scalar curvature G-invariants are constant on M.*

Before beginning the proof of this result, we must be just a bit more precise concerning foundational matters. Let $\pi : \mathcal{O}(\mathcal{M}) \to M$ be the principal $\mathrm{O}(n)$ bundle of orthonormal frames for the tangent bundle of M.

Definition 10.49

1. A *metric G-structure* on $\mathcal{M}$ is a reduction of $\mathcal{O}(\mathcal{M})$ to a principal G bundle $\mathcal{G}(\mathcal{M})$. Spin or Spin^c structures involve lifting the structure group and we will not consider those. A frame $u \in \mathcal{G}(\mathcal{M})$ will be called a *G-adapted basis*. Following Opozda [110], a linear isomorphism Φ from $T_P M$ to $T_Q M$ is said to be a *G-isometry* if Φ sends G-adapted bases for $T_P M$ to G-adapted

bases for $T_Q M$. If ϕ is a diffeomorphism from a small open neighborhood of P to a small open neighborhood of Q, then one says ϕ is a G-isometry if ϕ_* is a G-isometry from $T_{\tilde{P}} M$ to $T_{\phi \tilde{P}} M$ for all points $\tilde{P}$ in the domain of ϕ. The notion of *locally G-homogeneous* is defined analogously.

2. One can work purely algebraically. One says that $\mathcal{M}$ is *k-G-curvature homogeneous* if given any points P and Q of M, there exists a G-isometry Φ from $T_P M$ to $T_Q M$ so that $\Phi^* \nabla^k R_Q = \nabla^k R_P$. In fact, of course, it is only necessary to require this condition for $s \leq s(n, G)$ (where $s(n, G)$ is the Singer number), but we will ignore this point.

Example 10.50 If $G = U(\bar{n})$, then $\mathcal{G}(\mathcal{M})$ can be identified with the set of orthonormal frames for the complex tangent bundle; Φ is a G-isometry if and only if Φ is an isometry which commutes with the almost complex structure J, and ϕ is a G-isometry if and only ϕ is an isometry with $\phi_* J = J \phi_*$.

Since G is compact, any Lie sub-algebra of the Lie algebra of G is reductive and, consequently, by Opozda [110, Theorem 2.1], one has the following result.

Theorem 10.51 *A G-manifold $\mathcal{M}$ which is G k-curvature homogeneous for all k is locally G-homogeneous.*

The tangent bundle of M is the fiber bundle associated to $\mathcal{G}(\mathcal{M})$ under the natural action of G on $\mathbb{R}^n$, i.e., TM is the quotient space of $\mathcal{G}(\mathcal{M}) \times \mathbb{R}^n$ under the equivalence relation

$$(u, \xi) \sim (u \cdot g, g \cdot \xi) \quad \text{for} \quad g \in G \, .$$

We will denote the canonical projection from

$$\mathcal{G}(\mathcal{M}) \times \mathbb{R}^n \to TM$$

by $(u, \xi) \to u \cdot \xi$. In a system of local coordinates, we may represent $u = (e_1, \dots, e_n)$ where $\{e_i\}$ is an orthonormal basis for $T_P M$. If $\xi = (\xi^1, \dots, \xi^n) \in \mathbb{R}^n$, then $u \cdot \xi = \xi^i e_i \in T_P M$. The covariant derivative of order s of the curvature tensor with respect to an adapted basis u is given by $K^s(u) \in \otimes^{s+4}(\mathbb{R}^n)^*$; it is defined by

$$K^s(u)(\xi_1, \dots, \xi_{s+4}) := (\nabla^s R)_{\pi u}(u \cdot \xi_1, \dots, u \cdot \xi_{s+4}) \, .$$

Since G acts on $\mathbb{R}^m$ from the left, we have a dual action on $\otimes^{s+4}(\mathbb{R}^n)^*$ given by

$$(gT)(\xi_1, \dots, \xi_k) = T(g^* \xi_1, \dots, g^* \xi_k) \, .$$

By naturally, K^s is equivariant with respect to this action so $K^s(u \cdot g) = g^* K^s(u)$. Let

$$W_k := \oplus_{0 \leq s \leq k} \otimes^{s+4} (\mathbb{R}^n)^* \quad \text{and} \quad \Psi^k(u) := K^0(u) \oplus \cdots \oplus K^s(u) \, .$$

Let $\mathbb{R}[W_k]$ be the $\mathbb{R}$ unital polynomial algebra generated by $\mathbb{R}[W_m]$.

Definition 10.52 A polynomial $p \in \mathbb{R}[W_k]$ is said to be G-invariant if $p(g \cdot w) = p(w)$ for all $g \in G$ and all $w \in W_k$. Let $P[W_k]^G$ be the unital $\mathbb{R}$ algebra of G-invariant polynomials. For such a polynomial, the evaluation is independent of the particular adapted basis chosen and is called a *scalar curvature G-invariant.*

We have now introduced the necessary notation to establish Theorem 10.48. Since G is compact, it follows from the Weyl theory of invariants that $\mathbb{R}[W_m]^G$ is finitely generated (see the discussion in Procesi and Schwarz [122]). Moreover, if $\{p_1, \dots, p_t\}$ is a set of generators of $\mathbb{R}[W_k]^G$, the map $\hat{p} : W_k \to \mathbb{R}^t$, defined by $\hat{p}(w) = (p_1(w), \dots, p_t(w))$, separates orbits, that is, $\hat{p}(w) = \hat{p}(w')$ implies that $w' = aw$ for some $a \in G$.

Assume that all the scalar curvature G-invariants of M are constant. Let $f_1, \dots, f_t$ be the scalar curvature G-invariants associated with the invariant polynomials $p_1, \dots, p_t$. Then,

$$\hat{p}(\Phi^k(u)) = (p_1 \circ \Phi^k(u), \dots, p_t \circ \Phi^k(u)) = (f_1 \circ \pi(u), \dots, f_t \circ \pi(u)) = (f_1, \dots, f_t)$$

is constant. Since $\hat{p}$ separates orbits, $\Phi^k(\mathcal{G}(\mathcal{M}))$ is one single orbit of W_k. Therefore, for any two adapted bases u and v, there exists $g \in G$ such that $K^s(v) = K^s(ua^{-1})$ for any $s \geq 0$. Consequently, $\mathcal{M}$ is G-k-curvature homogeneous for every k. By Theorem 10.51, M is locally G-homogeneous. □

CHAPTER 11

Ricci Solitons

In this chapter, we will present some results concerning the geometry of Ricci solitons. We will focus on classification results which are related to suitably chosen geometric conditions. By working in the pseudo-Riemannian setting, we can construct Ricci solitons which do not have a Riemannian analogue. This is due, in part, to the existence of pseudo-Riemannian manifolds which are not flat and which admit non-trivial homothety vector fields.

Section 11.1 is an introduction to this material. We define the Hessian, Ricci solitons, gradient Ricci solitons, the Ricci flow, and other basic material for the convenience of the reader. In Section 11.2, we work in the Riemannian category and discuss the existence of homogeneous Ricci solitons and Ricci almost solitons paying special attention to the gradient case. We will then use work of Brozos-Vázquez et al. [15] and Fernández-López and García-Río [65] to extend a result of Petersen and Wylie [116] concerning the rigidity of homogeneous Riemannian gradient Ricci solitons to the broader class of curvature homogeneous manifolds.

Work of Brozos-Vázquez et al. [18] on locally homogeneous Lorentzian gradient Ricci solitons will be considered in Section 11.3. We will describe the structure of the Ricci tensor on a locally homogeneous Lorentzian gradient Ricci soliton. In the non-steady case, we show the soliton is rigid in dimensions 3 and 4. In the steady case, we give a complete classification in dimension 3. We then examine locally conformally flat gradient Ricci solitons. Since the gradient Ricci soliton Equation (11.1.b) only involves the Ricci curvature, it is natural to consider situations where the Ricci curvature determines the full curvature tensor. This is, of course, the case in dimension 3. We will examine the Riemannian case in Section 11.4 and the Lorentzian case in Section 11.5 separately. In Section 11.6, we discuss neutral signature self-dual gradient Ricci solitons.

11.1 INTRODUCTION

We begin by recalling some basic notational conventions.

Definition 11.1

1. Let $\rho(x, y) := \text{Trace}\{z \to \mathcal{R}(z, x)y\}$ be the *Ricci tensor* of an m-dimensional pseudo-Riemannian manifold $\mathcal{M} = (M, g)$. Let Ric be the associated *Ricci operator*; it is characterized by the identity $\rho(X, Y) = g(\text{Ric}\, X, Y)$. Let $\tau = \text{Tr}\{\text{Ric}\}$ be the *scalar curvature*. The *Ricci flow* is the solution to the equation $\partial_t g(t) = -2\rho_{g(t)}$.

2. The *Cotton tensor* is defined by setting
$\mathfrak{C}(X,Y,Z)=(\nabla_X\rho)(Y,Z)-(\nabla_Y\rho)(X,Z)-\frac{1}{2(m-1)}\{d\tau(X)g(Y,Z)-d\tau(Y)g(X,Z)\}$.
$\mathfrak{C}$ is conformally invariant in dimension 3 and vanishes if $\mathcal{M}$ is conformally flat.
3. The *Weyl tensor* is conformally invariant. It is defined for $m\geq 4$ by setting:
$W_{ijk\ell}:=R_{ijk\ell}+\frac{1}{m-2}(\rho_{i\ell}g_{jk}-\rho_{ik}g_{j\ell}+R_{jk}g_{i\ell}-R_{j\ell}g_{jk})$
$\quad+\frac{1}{(m-1)(m-2)}\tau(g_{ik}g_{j\ell}-g_{i\ell}g_{jk})$.
4. If $f\in C^\infty(M)$, let $\operatorname{Hess}_f(X,Y)=(\nabla_X df)(Y)=XY(f)-(\nabla_X Y)(f)$ be the *Hessian*. This is tensorial in (X,Y); $\operatorname{Hess}_f(\phi X,Y)=\operatorname{Hess}_f(X,\phi Y)=\phi\operatorname{Hess}_f(X,Y)$ for any smooth function ϕ. Since ∇ is torsion-free, $\operatorname{Hess}_f(X,Y)=\operatorname{Hess}_f(Y,X)$. In a system of local coordinates, the components of the Hessian are given by:
$$\{\operatorname{Hess}_f\}_{ij}=\partial_{x^i}\partial_{x^j}f-\Gamma_{ij}{}^k\partial_{x^k}f\,.$$
5. The *gradient vector field* ∇f is dual to the exterior derivative df; $g(\nabla f,X)=X(df)$.
6. The *Hessian operator* $\mathcal{H}_f(X):=\nabla_X(\nabla f)$ satisfies $\operatorname{Hess}(X,Y)=g(\mathcal{H}_f X,Y)$. We note that $\|\rho\|^2=\|\operatorname{Ric}\|^2$ and $\|\mathcal{H}_f\|^2=\|\operatorname{Hess}_f\|^2$.
7. Let $\mathcal{L}_X$ be the Lie derivative with respect to a vector field on M. The triple (M,g,X) is said to be a *Ricci soliton* if one has that
$$\mathcal{L}_X g+\rho=\lambda g\quad\text{for}\quad\lambda\in\mathbb{R}\,. \tag{11.1.a}$$
8. If $X=\nabla f$, then (M,g,f) is called a *gradient Ricci soliton* and the Ricci soliton Equation (11.1.a) takes the form
$$\operatorname{Hess}_f+\rho=\lambda\,g\,. \tag{11.1.b}$$
9. The Ricci soliton is said to be an *expanding soliton*, a *steady soliton* or a *shrinking soliton* if $\lambda<0$, $\lambda=0$, or $\lambda>0$, respectively.

Remark 11.2 We note that the Ricci tensor and the Hessian are well-defined in affine differential geometry since only the connection ∇ is involved. By contrast, the Ricci operator, the scalar curvature, the Hessian operator, and the gradient vector field are not well-defined in affine differential geometry since one needs the metric to raise or lower an index. Equation (11.1.b) is only well-defined in affine differential geometry if $\lambda=0$.

11.1.1 THE RICCI FLOW. One may set $X=0$ in the Ricci soliton Equation (11.1.a) or let f be constant in the gradient Ricci soliton Equation (11.1.b) to obtain the *Einstein equation* $\rho=\lambda g$. Consequently, Equation (11.1.a) and Equation (11.1.b) are natural generalizations of the Einstein equation and we will think of a Ricci soliton as a generalized Einstein manifold. The special significance of Ricci solitons comes from the analysis of the fixed points of the Ricci flow. The genuine fixed points of the Ricci flow are given by Ricci flat metrics. However, if (M, g_0) is an Einstein metric with constant $\lambda = \frac{\tau}{\dim(M)} \neq 0$, then $g(t) = (1 - 2\lambda t)g_0$ is a solution of the Ricci flow where the metric changes by homotheties. If $\lambda < 0$, then $g(t)$ is defined for $t > \frac{1}{2\lambda}$ and expands; if $\lambda > 0$, then $g(t)$ is defined for $t < \frac{1}{2\lambda}$ and shrinks.

We can consider the Ricci flow in the space of metrics modulo homotheties. If one looks for fixed points of the Ricci flow in this setting, then Einstein metrics are fixed points of the flow. More generally, one can also work modulo the action of the diffeomorphism group. A solution $g(t)$ of the Ricci flow is said to be *self-similar* if there exists a positive function $\sigma(t)$ and a one-parameter group of diffeomorphisms $\psi(t) : M \to M$ such that

$$g(t) = \sigma(t)\psi(t)^* g(0) . \tag{11.1.c}$$

Differentiating this relation yields $-2\,\rho_{g(t)} = \sigma'(t)\psi(t)^* g(0) + \sigma(t)\psi(t)^*(\mathcal{L}_Z g(0))$ where Z is the vector field given by $Z(\psi(t)(p)) = \frac{d}{dt}(\psi(t)(p))$ for any $p \in M$, and $\sigma' = \frac{d\sigma}{dt}$. Since one has $\rho_{g(t)} = \psi(t)^* \rho_{g(0)}$, one can drop the pullbacks in this relation to obtain

$$-2\,\rho_{g(0)} = \sigma'(t) g(0) + \mathcal{L}_{\tilde{Z}(t)} g(0) \quad \text{for} \quad \tilde{Z}(t) = \sigma(t) Z(t) .$$

Setting $\lambda = -\frac{1}{2}\dot{\sigma}(0)$ and $X = \frac{1}{2}\tilde{Z}(0)$, we see $-2\,\rho_{g(0)} = -2\lambda\, g(0) + 2\,\mathcal{L}_X g(0)$ at $t = 0$. This shows that for any self-similar solution of the Ricci flow there exists a vector field on M satisfying the Ricci soliton Equation (11.1.a).

Conversely, let X be a complete vector field on a pseudo-Riemannian manifold (M, g). Let $\psi(t) : M \to M$ be the 1-parameter family of diffeomorphisms with $\psi(0) = \mathrm{Id}$ satisfying the evolution equation $\partial_t \psi(t)(p) = (1 - 2\lambda t)^{-1} X(\psi(t)(p))$. We observe that $\psi(t)$ is defined for $t < (2\lambda)^{-1}$ if $\lambda > 0$ while if $\lambda < 0$, then $\psi(t)$ is defined for $t > (2\lambda)^{-1}$. The one-parameter family of metrics $g(t) := (1 - 2\lambda t)\psi(t)^* g$ satisfies

$$\partial_t g(t) = -2\lambda\, \psi(t)^* g + (1 - 2\lambda)\psi(t)^* \left(\mathcal{L}_{\frac{1}{1-2\lambda t} X} g\right) = \psi(t)^* \left(-2\lambda\, g + \mathcal{L}_{X(\psi(t)(p))} g\right) .$$

Hence, if the Ricci soliton Equation (11.1.a) holds, then

$$\partial_t\, g(t) = \psi(t)^*(-2\,\rho) = -2\,\psi(t)^* \rho = -2\,\rho(\psi(t)^* g) = -2\,\rho(g(t)) .$$

Consequently, $g(t)$ is a solution of the Ricci flow equation. It is an ancient solution in the shrinking case ($\lambda > 0$), an eternal solution in the steady case ($\lambda = 0$), and an immortal solution in the expanding case ($\lambda < 0$).

11.1.2 EXAMPLES OF NON-TRIVIAL GRADIENT RICCI SOLITONS. We refer to Cao [37] and Chow et al. [47] for more details concerning the material of this section.

Definition 11.3 Let $\mathcal{M} = (\mathbb{R}^2, g_0)$ where g_0 is the usual flat metric on $\mathbb{R}^2$. Since $\mathcal{M}$ is Ricci flat, $\mathcal{M}$ is a steady Ricci soliton. However, if we let $f(x) = \frac{\lambda}{2}\,\|x\|^2$, then $(\mathbb{R}^n, g_0, f)$ defines a gradient Ricci soliton which is expanding or shrinking depending on the sign of $\lambda \neq 0$. This soliton is known in the literature as the *Gaussian soliton*; it arises from the existence of homothety vector fields on Euclidean space which are not Killing vector fields. More generally, a gradient Ricci soliton (M, g, f) is said to be a *rigid soliton* if (M, g) is isometric to a quotient of $N \times \mathbb{R}^k$ where N is an Einstein manifold which has Einstein constant λ and the potential function $f = \frac{\lambda}{2}\|x\|^2$ is defined on the Euclidean factor. Although this is a rather restrictive condition, we will show subsequently that rigid Ricci solitons are the only examples in many important situations. We refer to Petersen and Wylie [117] for further details concerning rigid solitons.

Hamilton's cigar soliton is the steady gradient Ricci soliton given by the complete Riemann surface $\mathbb{R}^2$ equipped with the metric $g = \frac{1}{1+x^2+y^2}(dx^2 + dy^2)$ and the potential function $f(x, y) = -\log(x^2 + y^2 + 1)$. Note that this is also known in the Physics literature as *Witten's black hole*. Bryant [27] generalized Hamilton's cigar soliton to arbitrary dimensions by considering the Euclidean space in radial coordinates. Let g_{can} be the standard metric on the unit sphere S^{m-1} in $\mathbb{R}^m$ and let $(0, \infty) \times_\varphi S^{m-1}$ be $\mathbb{R}^m - \{0\}$ viewed as a warped product. The equations $f'' = (m+1)\frac{\varphi''}{\varphi}$ and $\varphi\varphi' f' = -m(1 - (\varphi')^2) + \varphi\,\varphi''$ specify a steady gradient Ricci soliton whose potential function only depends on the radial coordinate. To extend this smoothly thru the origin, one must examine the phase portrait corresponding to this system. We refer to Chow et al. [47] for details. There are many complete non-compact gradient Ricci solitons. The construction of Bryant was generalized by Dancer and Wang [52] and Ivey [82] who constructed expanding, steady, and shrinking gradient Ricci solitons on multiply warped products.

Two-distinct Ricci solitons (M, g, X_1) and (M, g, X_2) differ by a homothety vector field since $\mathcal{L}_{X_1 - X_2} g = (\lambda_1 - \lambda_2) g$. Thus, there exist non-Ricci flat pseudo-Riemannian manifolds which admit expanding, steady and shrinking Ricci solitons. For example, Cahen–Wallach symmetric spaces are steady gradient Ricci solitons (see Batat et al. [10]) as are plane waves (see Brozos-Vázquez, García-Río and Gavino-Fernández [17]). More generally, a 3-dimensional Lorentz Lie group admits a non-trivial Ricci soliton if and only if the Ricci operator has a single eigenvalue; note that the Ricci operator need not be diagonalizable in this setting. We refer to Brozos-Vázquez et al. [15] and Onda [109] for further information in the 3-dimensional setting. We also refer to Calvaruso and Fino [31] for examples of Ricci solitons on 4-dimensional non-reductive homogeneous spaces.

11.1.3 THE GRADIENT RICCI SOLITON EQUATION.

The gradient Ricci soliton Equation (11.1.b) relates geometric information of several different sorts. The Ricci tensor contains information related to the curvature of (M, g). The second fundamental form of level sets of f also plays a crucial role.

Definition 11.4 We say that a pseudo-Riemannian gradient Ricci soliton (M, g, f) is a *non-isotropic soliton* if $\|\nabla f\| \neq 0$. It is said to be an *isotropic soliton* if $\|\nabla f\| = 0$ and $\nabla f \neq 0$.

The following is a quite general result concerning gradient Ricci solitons in arbitrary signature. We refer to Chow et al. [47] and Petersen and Wylie [117] for further details.

Lemma 11.5 Let (M, g, f) be a pseudo-Riemannian gradient Ricci soliton.

1. $\nabla\tau = 2\operatorname{Ric}(\nabla f)$.
2. $\tau + \|\nabla f\|^2 - 2\lambda f = \text{const}$.
3. $R(X, Y, Z, \nabla f) = (\nabla_Y \rho)(X, Z) - (\nabla_X \rho)(Y, Z)$.
4. $\frac{1}{2}\Delta_f \tau = \lambda\tau - \|\rho\|^2$ where Δ_f is the f-Laplacian, $\Delta_f = \Delta - \nabla f$.
5. $(\nabla_{\nabla f}\operatorname{Ric}) + \operatorname{Ric}\circ\mathcal{H}_f = R(\nabla f, \cdot)\nabla f + \frac{1}{2}\nabla\nabla\tau$.

Proof. We take the trace of the gradient Ricci soliton Equation (11.1.b) to see that $\Delta f + \tau = m\lambda$. This implies that $\nabla\tau = -\nabla\Delta f$. Moreover, since $\nabla_Z\tau = 2\operatorname{div}(\operatorname{Ric}(Z))$ and $\operatorname{div}(\operatorname{Hess}_f)(X) = \rho(\nabla f, X) + g(\nabla\Delta f, X)$, one has

$$\begin{aligned} 0 &= \operatorname{div}(\lambda\, g)(X) = \operatorname{div}(\rho + \operatorname{Hess}_f)(X) \\ &= \tfrac{1}{2}\, g(X, \nabla\tau) + \rho(\nabla f, X) - g(\nabla\tau, X) = \rho(\nabla f, X) - \tfrac{1}{2}\, g(\nabla\tau, X)\,. \end{aligned}$$

Consequently, $\nabla\tau = 2\operatorname{Ric}(\nabla f)$ and Assertion 1 follows. Since $\nabla_{\nabla f}\nabla f = \frac{1}{2}\nabla\|\nabla f\|^2$,

$$\begin{aligned} 0 &= \rho(\nabla f, X) + \tfrac{1}{2}\, g(\nabla\Delta f, X) = \lambda\, g(X, \nabla f) - \operatorname{Hess}_f(X, \nabla f) + \tfrac{1}{2}\, g(\nabla\Delta f, X) \\ &= g(X, \lambda\nabla f - \nabla_{\nabla f}\nabla f + \tfrac{1}{2}\nabla\Delta f) = g(X, \lambda\nabla f - \tfrac{1}{2}\nabla\|\nabla f\|^2 - \tfrac{1}{2}\nabla\tau)\,. \end{aligned}$$

Consequently, $2\lambda f - \tau - \|\nabla f\|^2$ is constant and Assertion 2 follows. Assertion 3 follows from the identity $R(X, Y, \nabla f, Z) = (\nabla_Y H_f)(X, Z) - (\nabla_X H_f)(Y, Z)$, and from the gradient Ricci soliton Equation (11.1.b). We refer to Petersen and Wylie [117] for the proof of Assertion 4 and for the proof of the Assertion 5 in the Riemannian setting. One can use analytic continuation to extend Assertion 5 to the indefinite setting (or simply observe the proof goes through without change in the higher signature context, as a special case of the proof of Lemma 11.11). □

11.1.4 RICCI ALMOST SOLITONS. Pigola et al. [119] generalized the Ricci soliton Equation (11.1.a) by allowing λ to be a smooth function. In this setting, (M, g, X, λ) is said to be a *Ricci almost soliton*; the Ricci almost soliton is said to be a *proper Ricci almost soliton* if the soliton function λ is non-constant. In the special case when the vector field X is the gradient of a function f, Equation (11.1.b) becomes $\operatorname{Hess}_f + \rho = \lambda g$ after rescaling of the potential function f, and one refers to (M, g, f, λ) as a *gradient Ricci almost soliton.* If λ is non-constant, the gradient Ricci almost soliton is said to be a *proper gradient Ricci almost soliton*. There exist proper gradient Ricci almost solitons which correspond to self-similar solutions of some geometric flows. The *Ricci–Bourguignon flow* is given by the equation

$$\partial_t g(t) = -2\{\rho(t) - \kappa\tau(t)\, g(t)\} \quad \text{for} \quad \kappa \in \mathbb{R}\,.$$

This flow can be regarded as an interpolation between the Ricci flow and the Yamabe flow and corresponds to the equation $\frac{\partial}{\partial t} g(t) = -\tau(t)\, g(t)$. We refer to Catino et al. [40] for further details concerning the Ricci–Bourguignon flow. The self-similar solutions of this flow are called κ*-Einstein solitons* and correspond to the equation $\mathrm{Hess}_f + \rho = \{\kappa\, \tau + \mu\}\, g$ for some $\kappa, \mu \in \mathbb{R}$. They are gradient Ricci almost solitons with soliton function $\lambda = \kappa\, \tau + \mu$.

Not unexpectedly, Ricci almost solitons exhibit some similarities with but also striking differences from the usual Ricci solitons. For instance, no Kähler manifold admits proper gradient Ricci almost solitons (see Maschler [105, Proposition 3.1]). Furthermore, proper gradient Ricci almost solitons are irreducible (see Calviño-Louzao et al. [33, Lemma 2.4]). Finally, rigid gradient Ricci solitons are not irreducible.

11.2 RIEMANIAN HOMOGENEOUS RICCI ALMOST SOLITONS

11.2.1 ALGEBRAIC RICCI SOLITONS. As discussed in Section 11.1, Ricci solitons are fixed points of the Ricci flow modulo diffeomorphisms and rescaling. Automorphisms of a Lie group form a special kind of diffeomorphisms. Working modulo automorphisms of a Lie algebra leads to the notion of an *algebraic Ricci soliton* (see Lauret [100]).

Definition 11.6 Let (G, g) be a simply connected Lie group which is equipped with a left-invariant metric g. Let $\mathfrak{g}$ be the associated Lie algebra. If there exists $\lambda \in \mathbb{R}$ and if there exists a derivation D of the Lie algebra so that $\mathrm{Ric} = \lambda\, \mathrm{Id} + \mathrm{D}$, then g is said to be an *algebraic Ricci soliton.*

Any algebraic Ricci soliton on a non-Einstein Lie algebra gives rise to a Ricci soliton on the associated Lie group. So it is natural to wonder if there are non-algebraic Ricci solitons on Lie groups, and, more generally, on homogeneous spaces. This question was answered by Jablonski [83].

Theorem 11.7 *Let (M, g) be a homogeneous Ricci soliton. There exists a transitive group G of isometries such that $M = G/K$ is a G-algebraic Ricci soliton.*

There are many algebraic Ricci solitons (see, for example, the discussion in Will [135] and the references therein). The following result shows that the situation for Ricci almost solitons is quite different.

Theorem 11.8 *A locally homogeneous Riemannian manifold (M, g) admits a proper Ricci almost soliton if and only if (M, g) is either a space of constant sectional curvature or it is locally isometric to a product $\mathbb{R} \times N(c)$ where $N(c)$ is a space of constant curvature.*

Proof. Let (M, g, X, λ) be an n-dimensional proper Ricci almost soliton. If (M, g) is locally homogeneous, then for each point $p \in M$ there exist a local basis of Killing vector fields $\{\xi_1, \dots, \xi_n\}$. Take the Lie derivative with respect to ξ_i of the Ricci almost soliton equation. One may conclude that $X_\xi = [\xi, X]$ is a conformal vector field for each Killing vector field ξ on (M, g) by computing:

$$d\lambda(\xi_i)g = \mathcal{L}_{\xi_i}(\mathcal{L}_X g + \rho) = \mathcal{L}_{\xi_i}\mathcal{L}_X g = \mathcal{L}_X \mathcal{L}_{\xi_i} g + \mathcal{L}_{[\xi_i, X]} g = \mathcal{L}_{[\xi_i, X]} g \,.$$

Note that if all vector fields X_ξ vanish, then $d\lambda(\xi) = 0$ for all Killing vector fields. Therefore λ is constant due to the existence of local bases of Killing vector fields by local homogeneity. In this case the Ricci almost soliton becomes a Ricci soliton. Hence, we assume that there exists a Killing vector field ξ on (M, g) such that $X_\xi \neq 0$ is a non-Killing conformal vector field.

In the special case that $X_\xi = [\xi, X]$ is a homothety vector field (i.e., $d\lambda(\xi)$ is constant), fix a point $p \in M$ and use a basis of Killing vector fields $\{\xi_i\}$ to express $X_\xi(p) = \sum_\ell \kappa^\ell \xi_\ell(p)$. Then $Z = X_\xi - \sum_\ell \kappa^\ell \xi_\ell$ is a homothety vector field which vanishes at $p \in M$. Since homothety vector fields preserve the Ricci tensor ($\mathcal{L}_Z \rho = 0$) and $Z(p) = 0$, the eigenvalues of the Ricci operator vanish at $p \in M$. Indeed, let ρ_i be an eigenvalue of the Ricci operator with corresponding unit eigenvector e_i locally defined in a neighborhood of $p \in M$. Since Z is a homothety vector field, one has $\mathcal{L}_Z g = \mu g$, for some $\mu \in \mathbb{R}, \mu \neq 0$ and $\mathcal{L}_Z \rho = 0$. Then

$$\begin{aligned} 0 &= (\mathcal{L}_Z \rho)(e_i, e_i) = Z\rho(e_i, e_i) - 2\rho(\mathcal{L}_Z e_i, e_i) \\ &= Z(\rho_i) - 2\rho_i g(\mathcal{L}_Z e_i, e_i) = Z(\rho_i) + \mu\rho_i \,. \end{aligned}$$

It now follows that $\mu\rho_i(p) = Z_p(\rho_i) = 0$ since $Z_p = 0$. Hence, (M, g) is Ricci flat. Consequently, (M, g) is flat by local homogeneity (see Spiro [126, Theorem 3.4]).

The Weyl conformal curvature tensor W was defined in Definition 11.1. Assume X_ξ is a non-homothetic conformal vector field such that the divergence div $X_\xi = \frac{m}{2} d\lambda(\xi)$. Since X_ξ is conformal, its local flow consists of conformal transformations which preserve W. Consequently, $\mathcal{L}_{X_\xi} g = \frac{2}{m}\operatorname{div}(X_\xi) g$ and $\mathcal{L}_{X_\xi} W = 2\operatorname{div}(X_\xi) W$. Hence, we have

$$\begin{aligned} X_\xi \|W\|^2 &= X_\xi g(W, W) = (\mathcal{L}_{X_\xi} g)(W, W) + 2g(\mathcal{L}_{X_\xi} W, W) \\ &= d\lambda(\xi) g(W, W) + 2m d\lambda(\xi) g(W, W) = (2m + 1)\|W\|^2 d\lambda(\xi) \,. \end{aligned}$$

By local homogeneity, $\|W\|$ is constant on M. Consequently, $\|W\|^2 = 0$ since $d\lambda(\xi) \neq 0$. This shows that (M, g) is locally conformally flat (if $\dim(M) = m \geq 4$) and, therefore, locally symmetric by Takagi [130, Theorem A]. We remark here that although the original result of Takagi was stated for homogeneous manifolds, the proof remains valid for locally conformally flat spaces with constant Ricci curvatures. The 3-dimensional case is obtained in a completely analogous way considering the norm of the Cotton tensor (see Definition 11.1) instead of the Weyl tensor. Therefore (M, g) is either of constant sectional curvature, a product $\mathbb{R} \times N(c)$, or a product $N_1(c) \times N_2(-c)$ where $N(c)$ is a space of constant sectional curvature c. We complete the proof by considering these three possibilities. Since any space of constant curvature $N(c)$ is Einstein, Ricci almost solitons are conformal vector fields. This shows that $N(c)$ is a Ricci almost soliton for any value of $c \in \mathbb{R}$ (see, for example, Kanai [84, Theorem G]). The products $\mathbb{R} \times N(c)$ are rigid gradient Ricci solitons where we take the potential function $f = \frac{\mu}{2}(\pi_{\mathbb{R}})^2$. Here, $\pi_{\mathbb{R}}$ is the projection on $\mathbb{R}$ and the constant μ is determined by the sectional curvature $\mu = (m-1)c$. Let Z be a non-homothety conformal vector field on the universal cover of $\mathbb{R} \times N(c)$, i.e., $\mathcal{L}_Z g = \frac{2}{m}(\operatorname{div} Z)g$, with non-constant div Z. Then $X = Z + \frac{1}{2}\nabla f$ defines a Ricci almost soliton on (the universal cover of) $\mathbb{R} \times N(c)$ since

$$\mathcal{L}_X g + \rho = \mathcal{L}_Z g + \operatorname{Hess}_f + \rho = \left(\tfrac{2}{m} \operatorname{div} Z + \tfrac{m-1}{2} c\right) g\,.$$

The proof now follows since no proper Ricci almost solitons may occur in a product of the form $M = N_1(c) \times N_2(-c)$ where the factors have opposite constant sectional curvature (cf. Calviño-Louzao et al. [33, Lemma 2.2]). □

11.2.2 GRADIENT RICCI SOLITONS. The notion of a rigid soliton was defined in Definition 11.3. Petersen and Wylie [116] showed that homogeneous gradient Ricci solitons are rigid; this result was extended by Fernández-López and García-Río [65] to the more general setting of complete curvature homogeneous gradient Ricci solitons. If $\mathcal{M}$ is a curvature homogeneous pseudo-Riemannian manifold, then $\tau = \operatorname{Tr}\{\operatorname{Ric}\}$ and Rank(Ric) are constant.

Definition 11.9 Following Wang [133], we say that a non-constant function $f : M \to \mathbb{R}$ is a *transnormal function* if $\|\nabla f\|^2 = b(f)$ for some smooth function b on the range of f. The function f is said to be an *isoparametric function* if it also satisfies $\Delta f = a(f)$ for some function a on the range of f.

The following result extends Petersen and Wylie [116, Theorem 1] to the curvature homogeneous setting.

Theorem 11.10 *A complete gradient Ricci soliton with constant scalar curvature is rigid if and only if the Ricci operator has constant rank.*

Proof. Recall that a steady gradient Ricci soliton with constant scalar curvature is necessarily Ricci flat (see Petersen and Wylie [117]). Thus, we may assume that (M, g, f) is either expanding or shrinking. Assertion 2 in Lemma 11.5 then shows that, after a possible translation of the potential function, one has $\|\nabla f\|^2 = 2\lambda f - \tau$ and $\Delta f = m\lambda - \tau$. Thus the potential function is an isoparametric function on (M, g) with $a(f) = m\lambda - \tau$ and $b(f) = 2\lambda f - \tau$. If defined, let

$$f_- = \min_{x\in M} f(x), \quad f_+ = \max_{x\in M} f(x), \quad M_\pm = \{x \in M : f(x) = f_\pm\}\,.$$

The sets $M_\pm$ are called the the *focal varieties* of f and can be empty. If the soliton is shrinking, we will show that the potential function f has a minimum. Consequently, M_- is well-defined and non-empty. An analogous argument shows that if the soliton is expanding, then f has a maximum so M_+ is non-empty and well-defined. After translating f by a constant we assume that $2\lambda f = \|\nabla f\|^2$. This shows that f has the same sign as λ, and, moreover, it vanishes at the same points where ∇f does so. Hence, if $p \in M$ is such that $\nabla f(p) = 0$, then $f(p) = 0$ and the potential function attains its global minimum at $p \in M$. We therefore assume that $\nabla f \neq 0$ at every point and argue for a contradiction. Let $r := \{2f\lambda^{-1}\}^{1/2}$ map M to $\mathbb{R}$. One then has that $\nabla r = \frac{\sqrt{\lambda}}{2\|\nabla f\|}\nabla f$. Consequently, r satisfies the Eikonal equation $\|\nabla r\|^2 = 1$. Thus, r is a distance function and the integral curves of ∇r are geodesics. Since (M, g) is complete, this shows that ∇r is a complete vector field. So, if $\nabla f \neq 0$ then the range of r must be $\mathbb{R}$ which contradicts the fact that r is non-negative. This shows that M_- is non-empty in the shrinking setting.

Let f be a transnormal function (see Definition 11.9). Wang [133] showed that the focal varieties $M_\pm$ are smooth submanifolds of M if non-empty. Furthermore, the restriction of Hess_f to the focal varieties has only two eigenvalues, 0 and $\frac{1}{2}b'(f)$. Let $X, Y \in TM_\pm$ and $V, W \in TM_\pm^\perp$. One has $H_f(X,Y) = 0$ and $H_f(V,W) = \frac{1}{2}b'(f)g(V,W)$. Since the function b is linear with $b'(f) = 2\lambda$, the gradient Ricci soliton Equation (11.1.b) can be used to see that the restriction of the Ricci tensor to the (non-empty) focal submanifolds $M_\pm$ is of the form

$$\mathrm{Ric}_{|M_\pm} = \begin{pmatrix} \lambda I_k & 0 \\ 0 & 0_{n-k} \end{pmatrix} \quad \text{for} \quad k = \mathrm{Rank}(\mathrm{Ric}_{|M_\pm}) = \mathrm{codim}(M_\pm)\,.$$

It was shown by Petersen and Wylie [117, Proposition 1.3] that a gradient shrinking (resp. expanding) Ricci soliton with constant scalar curvature is rigid if and only if the Ricci tensor is bounded as $0 \leq \rho \leq \lambda g$ (resp. $\lambda g \leq \rho \leq 0$). Thus, to show that the Ricci soliton is rigid, it suffices to show that the Ricci tensor is bounded. By Lemma 11.5, the f-Laplacian of the scalar curvature satisfies the identity $\Delta_f \tau = \lambda\tau - \|\rho\|^2$. This shows that $\|\rho\|^2 = \lambda\tau$ in the constant scalar curvature setting (see, for example, Petersen and Wylie [117]). If one assumes that the rank of the Ricci operator is constant, i.e., $\mathrm{Rank}(\mathrm{Ric}) = \dim(M_\pm) = k$, then one has that the scalar curvature is $\tau = k\lambda$ since τ is constant. Let ρ_i, $i = 1, \dots, k$ be the non-zero eigenvalues of the Ricci operator at any point of M. Then

$$\sum_{i=1}^{k} (\rho_i - \lambda)^2 = \|\rho\|^2 - 2\lambda\tau + k\lambda^2 = \lambda\tau - 2\lambda\tau + \lambda\tau = 0\,.$$

This shows that all non-zero eigenvalues satisfy $\rho_1 = \rho_2 = \cdots = \rho_k = \lambda$. It now follows that Ricci operator has only two eigenvalues, 0 and λ, and, consequently, $0 \leq \rho \leq \lambda g$ in the shrinking case (resp. $\lambda g \leq \rho \leq 0$ in the expanding situation). □

The situation for gradient Ricci almost solitons is even more restrictive. We generalize the results in Lemma 11.5 to the setting at hand to become

Lemma 11.11 Let (M, g, f) be a gradient Ricci almost soliton.

1. $\Delta f + \tau = m\lambda$.
2. $\nabla\Delta f + \operatorname{Ric}(\nabla f) + \frac{1}{2}\nabla\tau = \nabla\lambda$.
3. $\nabla\tau = 2\operatorname{Ric}(\nabla f) + 2(m-1)\nabla\lambda$.
4. $R(X, Y, Z, \nabla f) = d\lambda(Y)g(X, Z) - d\lambda(X)g(Y, Z) + (\nabla_Y\rho)(X, Z) - (\nabla_X\rho)(Y, Z)$.
5. $\frac{1}{2}\Delta_f\tau = \lambda\tau - \|\rho\|^2 + (m-1)\Delta\lambda$.
6. $\nabla_{\nabla f}\operatorname{Ric} + \operatorname{Ric}\circ(\lambda\operatorname{Id} - \operatorname{Ric}) + g(\nabla\lambda, \cdot)\nabla f$
$= R(\nabla f, \cdot)\nabla f + g(\nabla f, \nabla\lambda)\operatorname{Id} + \nabla\cdot\nabla\left(\frac{\tau}{2} - (m-1)\lambda\right)$.

Proof. We refer to Calviño-Louzao et al. [33] and Pigola et al. [119] and the references therein for the proofs of the first 5 assertions. We sketch briefly the proof of Assertion 6. Note that $R(\nabla f, X)\nabla f = \nabla^2_{X,\nabla f}\nabla f - \nabla^2_{\nabla f, X}\nabla f$. We will compute both terms on the right-hand side. Since $\operatorname{Ric}(\nabla f) = \frac{1}{2}\nabla\tau - (m-1)\nabla\lambda$, we complete the proof by computing:

$$\begin{aligned}
\nabla^2_{X,\nabla f}\nabla f &= \nabla_X\nabla_{\nabla f}\nabla f - \nabla_{\nabla_X\nabla f}\nabla f = \nabla_X h_f(\nabla f) - h_f(\nabla_X\nabla f)\\
&= (\nabla_X h_f)(\nabla f) = (\nabla_X(\lambda\operatorname{Id} - \operatorname{Ric}))(\nabla f) = X(\lambda)\nabla f - (\nabla_X\operatorname{Ric})(\nabla f)\\
&= X(\lambda)\nabla f - \nabla_X\operatorname{Ric}(\nabla f) + \operatorname{Ric}(\nabla_X\nabla f)\\
&= X(\lambda)\nabla f - \nabla_X\operatorname{Ric}(\nabla f) + (\operatorname{Ric}\circ h_f)(X)\\
&= X(\lambda)\nabla f - \nabla_X\operatorname{Ric}(\nabla f) + (\operatorname{Ric}\circ(\lambda\operatorname{Id} - \operatorname{Ric}))(X)\\
&= X(\lambda)\nabla f - \tfrac{1}{2}\nabla_X\nabla\tau + (m-1)\nabla_X\nabla\lambda + (\operatorname{Ric}\circ(\lambda\operatorname{Id} - \operatorname{Ric}))(X),\\
\nabla^2_{\nabla f, X}\nabla f &= \nabla_{\nabla f}\nabla_X\nabla f - \nabla_{\nabla_{\nabla f}X}\nabla f = (\nabla_{\nabla f}h_f)X\\
&= (\nabla_{\nabla f}(\lambda\operatorname{Id} - \operatorname{Ric}))X = g(\nabla f, \nabla\lambda)X - (\nabla_{\nabla_f}\operatorname{Ric})X\,.
\end{aligned}$$

□

Theorem 11.12 *A curvature homogeneous Riemannian manifold (M, g) is a proper gradient Ricci almost soliton if and only if (M, g) is a space of non-zero constant sectional curvature.*

Proof. Let (M, g, f, λ) be a gradient Ricci almost soliton of dimension m. Assertion 6 of Lemma 11.11 implies that $g(\nabla\lambda, \cdot)\nabla f$ is symmetric and, therefore, $df \wedge d\lambda = 0$. Consequently, the soliton function λ is a function of the potential function f in a neighborhood of any point $p \in M$ where $df \neq 0$. For the remainder of the proof, we will set $\nabla\lambda = \varphi\nabla f$ where $\varphi = \pm\frac{\|\nabla\lambda\|}{\|\nabla f\|}$ is a smooth function in any open set where $df \neq 0$.

Since any curvature homogeneous manifold has constant Ricci curvatures, we may choose an orthonormal frame field which diagonalizes the Ricci operator. We then have

$$\operatorname{Ric} = \operatorname{diag}[\rho_1, \ldots, \rho_m]\,.$$

Since the scalar curvature is constant, Assertions 1 and 2 of Lemma 11.11 imply that one has $\operatorname{Ric}(\nabla f) = (1-m)\nabla\lambda$ and therefore $\operatorname{Ric}(\nabla f) = (1-m)\varphi\nabla f$. This shows that ∇f is an eigenvector of the Ricci operator. Since Ricci curvatures are constant, φ is constant and we may assume without loss of generality that $(1-m)\varphi = \rho_1$. This implies that $\Delta\lambda = \frac{\rho_1}{1-m}\Delta f$.

By Pigola et al. [119, Lemma 3.3], the weighted Laplacian of the scalar curvature of an m-dimensional gradient Ricci almost soliton (M, g, f, λ) satisfies

$$\tfrac{1}{2}\Delta_f \tau = \lambda\tau - \|\rho\|^2 + (m-1)\Delta\lambda\,.$$

This relationship together with Assertion 1 of Lemma 11.11 implies that

$$0 = \tfrac{1}{2}\Delta_f \tau = \lambda\tau - \|\rho\|^2 + (m-1)\Delta\lambda = \lambda\tau - \|\rho\|^2 - \rho_1(m\lambda - \tau)\,.$$

Therefore, a gradient Ricci almost soliton with constant Ricci curvatures is a gradient Ricci soliton if $\tau \neq m\rho_1$. Moreover, if $\tau = m\rho_1$, then $\|\rho\|^2 = \rho_1\tau = m\rho_1^2$. One may then use the Cauchy–Schwarz inequality to see that $\|\rho\|^2 \geq \frac{\tau^2}{m} = m\rho_1^2$. Thus $\|\rho\|^2 = \frac{\tau^2}{m}$. Consequently, any proper gradient Ricci almost soliton with constant Ricci curvatures is Einstein. □

11.3 LORENTZIAN HOMOGENEOUS GRADIENT RICCI SOLITONS

Let (M, g) be a Lorentzian manifold of dimension $m + 2 \geq 3$. We will assume for the most part that (M, g) is locally homogeneous and, consequently, the scalar curvature is constant.

One has canonical examples which play a central role in the theory. Let (N, g_N) be an Einstein manifold with *Einstein constant* λ, i.e., $\rho_N = \lambda\, g_N$. Let $M = N \times \mathbb{R}^k$ have the product metric g_M and let $f(x) := \frac{\lambda}{2}\|\pi(x)\|^2$ where π is projection on the second factor. Then (M, g_M, f) is a gradient Ricci soliton and is said to be *rigid*. Since we are interested in questions of local geometry, by an abuse of notation we will also say that (M, g_M, f) is *rigid* if (M, g_M, f) is isomorphic to an open subset of a product $N \times \mathbb{R}^k$ which is rigid. We will use the following results of Petersen and Wylie [116]. Assertion 2 was first proved in the Riemannian setting but extends easily to arbitrary signature.

Theorem 11.13

1. *Any locally homogeneous Riemannian gradient Ricci soliton is rigid.*
2. *Let $(M, g) = (M_1 \times M_2, g_1 \oplus g_2)$ be the direct product of two pseudo-Riemannian manifolds. If f satisfies the gradient Ricci soliton equation on (M, g), then we have that $f(x_1 + x_2) = f_1(x_1) + f_2(x_2)$ where f_1 and f_2 satisfy the gradient Ricci soliton equation on (M_1, g_1) and on (M_2, g_2) separately.*

Assertion 1 was originally proven for homogeneous manifolds, but the assumption of homogeneity can be weakened to local homogeneity by modifying the argument in Petersen and Wylie [116, Proposition 1] as in the proof of Assertion 2c of Lemma 11.14. Since any locally homogeneous Riemannian gradient Ricci soliton is rigid, the classification is complete in this context. However the possible geometries are much richer in the Lorentzian setting owing to the existence of degenerate parallel line fields. For example, in Theorem 11.24, we will present results of Batat et al. [10] showing that Cahen–Wallach spaces admit steady non-rigid gradient Ricci solitons.

For the remainder of this section, we will assume (unless otherwise noted) that the underlying manifold (M, g) is a locally homogeneous Lorentzian manifold and that (M, g, f) is a gradient Ricci soliton. In Theorems 11.16–11.19 we present results concerning non-steady solitons ($\lambda \neq 0$). In low dimensions, such solitons are rigid; in arbitrary dimensions, the eigenvalue structure of the Ricci operator agrees with the corresponding eigenvalue structure of a rigid soliton, i.e., there are only two eigenvalues $\{0, \lambda\}$. In Theorems 11.21–11.22, we present our results concerning steady solitons ($\lambda = 0$). Theorem 11.21 gives a complete classification if $\|\nabla f\|^2 < 0$. In Theorem 11.22, we will examine the situation when $\|\nabla f\|^2 = 0$ and show the Ricci tensor is either 2-step or 3-step nilpotent; it follows from work of Leistner and Nurowski [102] that the metrics in question are pure radiation metrics with parallel rays. If we further restrict the geometry, stronger results are available. We give a complete classification of symmetric Lorentzian gradient Ricci solitons in Theorem 11.25. In Theorem 11.28, we give a complete classification of 3-dimensional Lorentzian locally homogeneous gradient Ricci solitons; there are three nontrivial families of examples.

The fact that (M, g) is Lorentzian plays a crucial role in many arguments. For example, when we study the non-steady case, there exists a distinguished null parallel vector field and there do not exist orthogonal null vector fields. This is a Lorentzian phenomena not present in the Riemannian or the higher signature setting. The fact that (M, g) is locally homogeneous is not simply used to ensure that the scalar curvature is constant; it plays a role in many proofs where we take frame fields consisting at least in part of Killing vector fields. As our discussion is local in nature, it is not necessary to impose global conditions such as global homogeneity or completeness.

11.3.1 GRADIENT RICCI SOLITONS WITH τ CONSTANT. The following is a quite general result concerning gradient Ricci solitons with constant scalar curvature τ in arbitrary signature.

Lemma 11.14 Let (M, g, f) be a gradient Ricci soliton with constant scalar curvature.

1. We have the following relations:
   (a) $\mathrm{Ric}(\nabla f) = 0$.
   (b) $\|\nabla f\|^2 - 2\lambda f = \text{const}$.
   (c) $R(X, Y, \nabla f, Z) = (\nabla_X \rho)(Y, Z) - (\nabla_Y \rho)(X, Z)$.
   (d) $(\nabla_{\nabla f} \mathrm{Ric}) + \mathrm{Ric} \circ \mathcal{H}_f = \mathcal{R}(\nabla f, \cdot)\nabla f$.
2. Let X be a Killing vector field.
   (a) $\mathcal{L}_X (\mathrm{Hess}_f) = \mathrm{Hess}_{X(f)}$.

(b) $\nabla\{X(f)\}$ is a parallel vector field.

(c) If $\lambda \neq 0$, then $\nabla\{X(f)\} = 0$ if and only if $X(f) = 0$.

3. $\lambda((m+2)\lambda - \tau) = \| \operatorname{Hess}_f \|^2$.
4. If (M, g, f) is isotropic and non-steady, then (M, g) is an Einstein manifold.
5. If (M, g, f) is steady, then $\| \operatorname{Hess} f \|^2 = 0$ and $\|\nabla f\|^2 = \mu$ is constant.

Proof. Assertion 1 follows from Lemma 11.5 assuming $\nabla\tau = 0$. Let X be a Killing vector field. Choose a point P of M so that $X(P) \neq 0$; Assertion 2 for P where $X(P) = 0$ will then follow by continuity. Choose a system of local coordinates $(x^1, \dots, x^{m+2})$ so that $X = \partial_{x^1}$. Set $g_{ij} := g(\partial_{x^i}, \partial_{x^j})$ and observe that

$$\begin{aligned} \partial_{x^1} g_{ij} &= g(\nabla_{\partial_{x^1}} \partial_{x^i}, \partial_{x^j}) + g(\partial_{x^i}, \nabla_{\partial_{x^1}} \partial_{x^j}) \\ &= g(\nabla_{\partial_{x^i}} \partial_{x^1}, \partial_{x^j}) + g(\partial_{x^i}, \nabla_{\partial_{x^j}} \partial_{x^1}) = (\mathcal{L}_{\partial_{x^1}} g)(\partial_{x^i}, \partial_{x^j}) . \end{aligned}$$

Thus, $\partial_{x^1} g_{ij} = 0$ so $\partial_{x^1} \Gamma_{ij}{}^k = 0$ as well. We establish Assertion 2-a by computing:

$$\begin{aligned} (\mathcal{L}_{\partial_{x^1}} \operatorname{Hess}_f)(\partial_{x^i}, \partial_{x^j}) &= \mathcal{L}_{\partial_{x^1}} \operatorname{Hess}_f(\partial_{x^i}, \partial_{x^j}) \\ &= \mathcal{L}_{\partial_{x^1}} \left(\partial^2_{x^i x^j}(f) - \Gamma_{ij}{}^k \partial_{x^k}(f) \right) \\ &= \partial^3_{x^1 x^i x^j}(f) - \partial_{x^1}(\Gamma_{ij}{}^k)\partial_{x^k}(f) - \Gamma_{ij}{}^k \partial^2_{x^1 x^k}(f) \\ &= \partial^2_{x^i x^j} \partial_{x^1}(f) - \Gamma_{ij}{}^k \partial_{x^k} \partial_{x^1}(f) = \operatorname{Hess}_{\partial_{x^1}(f)}(\partial_{x^i}, \partial_{x^j}) . \end{aligned}$$

Because $\mathcal{L}_X g = 0$ and since ρ is natural, one has that $\mathcal{L}_X \rho = 0$. Equation (11.1.b) implies that $\mathcal{L}_X \operatorname{Hess}_f = 0$, and therefore by Assertion 2-a, $\operatorname{Hess}_{X(f)} = 0$. Consequently, $\nabla\{X(f)\}$ is parallel. This establishes Assertion 2-b. Assume now that $\lambda \neq 0$. It is clear that $\nabla\{X(f)\} = 0$ if $X(f) = 0$. Conversely, if $\nabla\{X(f)\} = 0$, then $X(f) = \kappa$ for some constant κ. Since the scalar curvature is constant, Assertion 1 implies that $\operatorname{Ric}(\nabla f) = 0$. Since X is a Killing vector field,

$$\begin{aligned} 0 &= \nabla f(\kappa) = \nabla f(X(f)) = \nabla f\, g(\nabla f, X) = g(\nabla_{\nabla f} \nabla f, X) + g(\nabla f, \nabla_{\nabla f} X) \\ &= \operatorname{Hess}_f(\nabla f, X) + \tfrac{1}{2}(\mathcal{L}_X g)(\nabla f, \nabla f) = -\rho(\nabla f, X) + \lambda\, g(\nabla f, X) = \lambda \kappa . \end{aligned}$$

Thus, $\kappa = 0$. Consequently, $\nabla\{X(f)\} = 0$ if and only if $X(f) = 0$. This establishes Assertion 2-c.

Next, we turn our attention to Assertion 3. We have the *Bochner identity*:

$$\tfrac{1}{2} \Delta g(\nabla f, \nabla f) = \| \operatorname{Hess}_f \|^2 + \rho(\nabla f, \nabla f) + g(\nabla \Delta f, \nabla f) . \tag{11.3.a}$$

By Assertion 1, $\operatorname{Ric}(\nabla f) = 0$ and $\|\nabla f\|^2 - 2\lambda f = \text{const}$. Thus, the left-hand side of Equation (11.3.a) becomes $\frac{1}{2} \Delta g(\nabla f, \nabla f) = \lambda \Delta f - \frac{1}{2} \Delta\tau$. Taking the trace in Equation (11.1.b) shows that $\Delta f = (m+2)\lambda - \tau$ and, therefore, $\frac{1}{2} \Delta g(\nabla f, \nabla f) = \lambda((m+2)\lambda - \tau)$. On the other hand, since $\mathrm{Ric}(\nabla f) = 0$ and $\nabla \Delta f = -\nabla \tau = 0$, the right-hand side in Bochner formula reduces to $\| \mathrm{Hess}_f \|^2$. This completes the proof of Assertion 3.

If $\|\nabla f\|^2 = 0$, we may apply Assertion 1 to see $2\lambda f = \mathrm{const}$. Since $\lambda \neq 0$, f is constant and (M, g) is an Einstein manifold. Assertion 4 follows. Finally, if $\lambda = 0$, then one has that $\| \mathrm{Hess}_f \|^2 = 0$. By Equation (11.1.b), $\mathcal{H}_f = -\mathrm{Ric}$. Consequently, $\mathrm{Ric}(\nabla f) = 0$ implies that $\mathcal{H}_f(\nabla f) = 0$. Therefore, ∇f is a geodesic vector field. Since $\tau + \|\nabla f\|^2 - 2\lambda f = \mathrm{const}$, $\|\nabla f\|^2$ is constant and therefore f is a solution of the *Eikonal equation* $\|\nabla f\|^2 = \mu$. This establishes Assertion 5 and completes the proof. □

Setting $\lambda \neq 0$ or $\lambda = 0$ in Lemma 11.14 gives significantly different information about the potential function f. Consequently, different techniques are required to study the steady setting from those which are required to study the non-steady setting. By Lemma 11.14, any isotropic non-steady gradient Ricci soliton with constant scalar curvature is an Einstein manifold. However, there exist isotropic steady gradient Ricci solitons which are not Einstein (see, for example, Batat et al. [10]).

11.3.2 NON-STEADY LOCALLY HOMOGENEOUS LORENTZIAN GRADIENT RICCI SOLITONS.

Definition 11.15 We say that a Lorentzian manifold (M, g) is *irreducible* if the holonomy representation has no non-trivial invariant subspace. We say that (M, g) is *indecomposable* if the metric on any non-trivial subspace fixed by the holonomy representation is degenerate. Consequently the holonomy representation does not decompose as a non-trivial direct sum of subrepresentations. This distinction is only relevant in the indefinite setting; any Riemannian manifold is irreducible if and only if it is indecomposable.

Let $\mathbb{R}^k_\nu$ denote $\mathbb{R}^k$ with the usual flat Riemannian metric if $\nu = 0$ (Euclidean space) or the usual flat Lorentzian metric if $\nu = 1$ (Minkowski space).

Theorem 11.16 *Let (M, g, f) be a locally homogeneous Lorentzian non-steady gradient Ricci soliton. Then one of the following holds.*

1. *(M, g) is an irreducible Einstein manifold.*
2. *(M, g, f) is rigid. There is a local splitting $(M, g, f) = (N \times \mathbb{R}^k_\nu, g_N + g_e, f_N + f_e)$ where (N, g_N) is an Einstein manifold with Einstein constant λ, $f_e(x) := \frac{\lambda}{2}\|x\|^2$, and $\nu = 0, 1$.*
3. *(M, g, f) locally splits as $(M, g, f) = (N_0 \times N_1 \times \mathbb{R}^k, g_0 + g_1 + g_e, f_0 + f_1 + f_e)$ where (N_0, g_0, f_0) is an indecomposable locally homogeneous Lorentzian gradient Ricci soliton, (N_1, g_1) is a Riemannian Einstein manifold with Einstein constant λ, and $(\mathbb{R}^k, g_e, f_e)$ is Euclidean space with $f_e(x) := \frac{\lambda}{2}\|x\|^2$.*

Proof. Assume that (M, g) is irreducible or, equivalently, that there are no non-trivial parallel distributions on M. Consequently, any parallel vector field is trivial. Let X be a Killing vector field. Because $\nabla\{X(f)\}$ is a parallel vector field, $\nabla\{X(f)\} = 0$. This implies that $X(f)$ is constant and, therefore, $X(f) = 0$. Since the underlying Lorentzian structure (M, g) is locally homogeneous, there are $m + 2$ linearly independent Killing vector fields $X_1, \dots, X_{m+2}$ locally defined. Consequently, f is constant and the metric is Einstein. This establishes Assertion 1 of Theorem 11.16.

We now apply the local splitting result of Assertion 2 of Theorem 11.13. Let X be a Killing vector field on (M, g). If $\nabla\{X(f)\}$ is spacelike or timelike, then we may split, at least locally, a 1-dimensional factor from (M, g) and decompose locally

$$(M, g, f) = (N \times \mathbb{R}, g_N \oplus g_e, f_N + f_e)\,.$$

If $\nabla\{X(f)\}$ is timelike, then (N, g_N) is Riemannian and by Assertion 1 of Theorem 11.13, rigid, which would finish the discussion. Thus, we may assume (N, g_N) is Lorentzian so $\nabla\{X(f)\}$ is spacelike and the factor $(\mathbb{R}, g_e)$ is positive definite. We proceed inductively to decompose $(M, g, f) = (N \times \mathbb{R}^k, g_N \oplus g_e, f_N + f_e)$ (at least locally) so that (N, g_N, f_N) is a locally homogeneous Lorentzian Ricci soliton with $\nabla\{X(f)\}$ null or zero for all Killing vector fields X. Now two possibilities may occur. If N is indecomposable, Assertion 3 follows with trivial N_1. If N is decomposable, then either N is an Einstein manifold and Assertion 2 holds (this is the case if $\nabla\{X(f)\} = 0$ for all Killing vector fields in N) or N decomposes in the form $N = N_0 \times N_1$ where N_0 is Lorentzian and indecomposable (the latter happens if there exists a Killing vector field X so that $\nabla\{X(f)\}$ is null). (N_1, g_1, f_1) is a Riemannian locally homogeneous gradient Ricci soliton which, as a consequence of Theorem 11.13, is an Einstein manifold. □

We now examine the situation in Assertion 3 of Theorem 11.16 and study the indecomposable factor. Recall that a Lorentzian manifold is said to be a *Walker manifold* if it admits a parallel null line field, and a *strict Walker manifold* if this distribution is spanned by a parallel null vector field; we refer to Brozos-Vázquez et al. [21] for further details.

Definition 11.17 Let $\mathfrak{S} := \rho - \frac{\tau}{2(m+1)} g$ be the *Schouten tensor* $\mathfrak{S}$. We will say that (M, g) has a *harmonic Weyl tensor* if the Schouten tensor is *Codazzi*. This means (see Besse [13]) that $\nabla_X \mathfrak{S}_{YZ} = \nabla_Y \mathfrak{S}_{XZ}$.

Theorem 11.18 *Let (M, g, f) be a locally homogeneous indecomposable Lorentzian non-steady gradient Ricci soliton which is not an Einstein manifold.*

1. *Locally, there exists a Killing vector field X so $U := \nabla\{X(f)\}$ is a non-trivial parallel null vector field. Therefore, (M, g) is a strict Walker manifold.*
2. *U is unique up to scale, $\mathcal{V} := \{U, \nabla f\} \subset \ker\{\mathrm{Ric}\}$ is a U-parallel Lorentzian distribution, and $\nabla\{U(f)\} = \lambda U$.*

3. $\nabla_U \text{Ric} = \nabla_U \mathcal{H}_f = 0$, $\text{Spec}\{\text{Ric}\} = \text{Spec}\{\mathcal{H}_f\} = \{0, \lambda\}$, Ric *and* $\mathcal{H}_f$ *are diagonalizable,* $\ker\{\text{Ric}\} = \text{Image}\{\mathcal{H}_f\}$, *and* $\ker\{\mathcal{H}_f\} = \text{Image}\{\text{Ric}\}$.
4. *The Weyl tensor of* (M, g) *is harmonic if and only if* (M, g, f) *is rigid.*
5. *If* $\dim(\ker\{\text{Ric}\}) = 2$, *then* (M, g, f) *is rigid.*

Proof. We establish the assertions of Theorem 11.18 seriatim. First suppose that (M, g) is neither decomposable nor Einstein. To prove Assertion 1, we must show there exists X so $U = \nabla\{X(f)\}$ is a parallel null vector field. Let Z be any Killing vector field. Since (M, g) is not decomposable and since $\nabla\{Z(f)\}$ is parallel, $\nabla\{Z(f)\}$ must be isotropic. If $\nabla\{Z(f)\}$ vanishes for all such Z, then f is constant and, therefore, (M, g) is an Einstein manifold which is contrary to our assumption. Thus, $U := \nabla\{Z(f)\}$ has the desired properties for some Killing vector field Z.

To establish Assertion 2, we must show that U is unique up to scale, that U belongs to $\ker\{\text{Ric}\}$, and that $\nabla\{U(f)\} = \lambda U$. Suppose that there are two Killing vector fields Z_1 and Z_2 on (M, g) so that $\nabla\{Z_1(f)\}$ and $\nabla\{Z_2(f)\}$ are linearly independent. Since the signature is Lorentzian, $\text{span}\{\nabla\{Z_1(f)\}, \nabla\{Z_2(f)\}\}$ cannot be a null distribution. Consequently, there exists a linear combination $Z = a_1 Z_1 + a_2 Z_2$ so $\nabla\{Z(f)\}$ is either timelike or spacelike. This implies that (M, g) is decomposable which is false. Thus, the vector field $U = \nabla\{Z(f)\}$ is unique up to scale.

Since U is parallel, it is a Killing vector field and, therefore, $\nabla\{U(f)\} = \alpha U$ for some $\alpha \in \mathbb{R}$. We must now show that $\text{Ric}(U) = 0$. Let $\{Z_1, Z_2, \dots, Z_{m+2}\}$ be a local basis of Killing vector fields. Choose the notation so $Z = Z_1$. We then have $\nabla\{Z_i(f)\} = \mu_i U$ for $i \geq 2$. Since $\nabla\{Z_i(f)\}$ is parallel, necessarily μ_i is constant. By replacing Z_i by $Z_i - \mu_i Z_1$, we may assume therefore that $\nabla\{Z_i(f)\} = 0$ for $i \geq 2$. Since $\lambda \neq 0$, Lemma 11.14 shows $Z_i(f) = 0$ for $i \geq 2$. We use Equation (11.1.b) and the fact that $g(\nabla f, X) = X(f)$ to see

$$\begin{aligned} g(U, \nabla f) &= g(\nabla\{Z_1(f)\}, \nabla f) = g(\nabla\{g(Z_1, \nabla f)\}, \nabla f) \\ &= \nabla f\, g(Z_1, \nabla f) = g(\nabla_{\nabla f} Z_1, \nabla f) + g(Z_1, \nabla_{\nabla f} \nabla f) \\ &= \text{Hess}_f(Z_1, \nabla f) = \lambda g(Z_1, \nabla f) = \lambda Z_1(f) \neq 0\,. \end{aligned}$$

By Lemma 9.49, $g(\nabla_{\nabla f} Z_1, \nabla f) = 0$ since Z_1 is a Killing vector field. As $g(U, \nabla f) \neq 0$ and as U is a null vector, $\mathcal{V} := \text{span}\{U, \nabla f\}$ has Lorentzian signature. By Lemma 11.14, $\nabla\{U(f)\} \neq 0$ so $\alpha \neq 0$.

If X is an arbitrary vector field, we study $\mathcal{H}_f(U)$ by computing:

$$\text{Hess}_f(X, U) = g(U, \nabla_X \nabla f) = X g(U, \nabla f) = g(X, \nabla\{U(f)\}) = \alpha g(X, U)\,.$$

Thus, $\mathcal{H}_f(U) = \alpha U$. As $\mathcal{H}_f(\nabla f) = \lambda \nabla f$, $\alpha g(\nabla f, U) = \text{Hess}_f(\nabla f, U) = \lambda g(\nabla f, U)$. This shows that $\alpha = \lambda$. By Equation (11.1.b), $\text{Ric}(U) = 0$. Since $\nabla_U U = 0$ and since $\nabla_U \nabla f = \lambda U$, ∇_U preserves $\mathcal{V} \subset \ker\{\text{Ric}\}$. This proves Assertion 2.

We showed the Lorentzian distribution $\mathcal{V} := \text{span}\{U, V\} \subset \ker\{\text{Ric}\}$ is U-parallel. Consequently, $\mathcal{V}^\perp$ is a Ric-invariant distribution with a positive definite signature. Since Ric is self-adjoint, there exists an orthonormal basis $\{E_1, \ldots, E_m\}$ of $\mathcal{V}^\perp$ so that $\mathrm{Ric}(E_i) = \alpha_i E_i$. Since (M, g) is locally homogeneous, the coefficient functions α_i are constant for one has $1 \leq i \leq m$. This proves in particular that Ric and $\mathcal{H}_f = \lambda\,\mathrm{Id} - \mathrm{Ric}$ are diagonalizable. We now show that ∇_U preserves the eigenspaces in $\mathcal{V}^\perp$. For $i \neq j$, since U is parallel, $R(U, E_i, E_j, \nabla f) = 0$. By Assertion 1 of Lemma 11.14 we have:

$$\begin{aligned}
0 = R(U, E_i, \nabla f, E_j) &= (\nabla_U \rho)(E_i, E_j) - (\nabla_{E_i} \rho)(U, E_j),\\
U &= \rho(E_i, E_j) - \rho(\nabla_U E_i, E_j) - \rho(E_i, \nabla_U E_j)\\
&\quad - E_i \rho(U, E_j) + \rho(\nabla_{E_i} U, E_j) + \rho(U, \nabla_{E_i} E_j)\\
&= -\alpha_j g(\nabla_U E_i, E_j) - \alpha_i g(E_i, \nabla_U E_j)\\
&= (\alpha_i - \alpha_j) g(\nabla_U E_i, E_j).
\end{aligned}$$

This shows that if E_i and E_j belong to different eigenspaces, then $\nabla_U E_i$ is orthogonal to E_j. Hence, ∇_U commutes with Ric and, as a consequence of the Ricci soliton Equation (11.1.b) it also commutes with $\mathcal{H}_f$. Consequently, as desired, $\nabla_U \mathrm{Ric} = 0$ and $\nabla_U \mathcal{H}_f = 0$.

We must show that 0 and λ are the only eigenvalues of Ric. Normalize V to be a multiple of ∇f so $g(V, V) = \epsilon = \pm 1$. Let S be any level set of f. The integral curves of U are transversal to S because $g(U, \nabla f) \neq 0$. Use parallel transport along the integral curves of U to extend the local frame $\{E_1, \ldots, E_m\}$ from S to a neighborhood of S to define a local frame field $\{F_1, \ldots, F_m\}$ for $\mathcal{V}^\perp$ such that $\nabla_U F_i = 0$. Since $\nabla_U \mathrm{Ric} = 0$, the vector fields F_i are still eigenvectors of the Ricci operator Ric. We will use this local frame field to see that Ric has only two eigenvalues $\{0, \lambda\}$. First note that

$$\begin{aligned}
& (\nabla_{\nabla f} \rho)(F_i, F_i) = \nabla f\, \rho(F_i, F_i) - 2\rho(\nabla_{\nabla f} F_i, F_i)\\
= \;& \alpha_i \nabla f\, g(F_i, F_i) - 2\alpha_i g(\nabla_{\nabla f} F_i, F_i) = \alpha_i \left(\nabla_{\nabla f} g\right)(F_i, F_i) = 0.
\end{aligned}$$

We use Lemma 11.14 to compute:

$$\begin{aligned}
\rho(F_i, F_i) &= \epsilon R(F_i, V, F_i, V)\\
&\quad + \textstyle\sum_{j\neq i} R(F_i, F_j, F_i, V) g(F_j, V) + \sum_{j\neq i} R(F_i, F_j, F_i, F_j)\\
&= \tfrac{\epsilon}{\|\nabla f\|^2} \left((\nabla_{F_i} \rho)(\nabla f, F_i) - (\nabla_{\nabla f} \rho)(F_i, F_i)\right)\\
&\quad + \textstyle\sum_{j\neq i} R(F_i, F_j, F_i, V) g(F_j, V) + \sum_{j\neq i} R(F_i, F_j, F_i, F_j)\\
&= \tfrac{\epsilon}{\|\nabla f\|^2} \left(F_i \rho(\nabla f, F_i) - \rho(\nabla_{F_i} \nabla f, F_i) - \rho(\nabla f, \nabla_{F_i} F_i)\right)\\
&\quad + \textstyle\sum_{j\neq i} R(F_i, F_j, F_i, V) g(F_j, V) + \sum_{j\neq i} R(F_i, F_j, F_i, F_j)\\
&= -\tfrac{\epsilon}{\|\nabla f\|^2} \rho(\mathcal{H}_f F_i, F_i)\\
&\quad + \textstyle\sum_{j\neq i} R(F_i, F_j, F_i, V) g(F_j, V) + \sum_{j\neq i} R(F_i, F_j, F_i, F_j).
\end{aligned}$$

Because $\nabla_U \rho = 0$, $U\rho(F_i, F_i) = 2\rho(\nabla_U F_i, F_i) = 0$. We now differentiate the three summands in the previous expression with respect to U:

$$
\begin{aligned}
U(-\tfrac{1}{\|\nabla f\|^2}\rho(\mathcal{H}_f F_i, F_i)) &= \tfrac{Ug(\nabla f,\nabla f)}{\|\nabla f\|^4}\rho(\mathcal{H}_f F_i, F_i) - \tfrac{1}{\|\nabla f\|}U\rho(\mathcal{H}_f F_i, F_i)\\
&= \tfrac{2\lambda g(U,\nabla f)}{\|\nabla f\|^4}\rho(\mathcal{H}_f F_i, F_i) - \tfrac{1}{\|\nabla f\|}\left(\rho(\nabla_U \mathcal{H}_f F_i, F_i) + \rho(\nabla_{F_i}\nabla f, \nabla_U F_i)\right)\\
&= \tfrac{2\lambda g(U,\nabla f)}{\|\nabla f\|^4}\rho(\mathcal{H}_f F_i, F_i) - \tfrac{1}{\|\nabla f\|}\left(\rho(\mathcal{H}_f(\nabla_U F_i), F_i) + \rho(\nabla_{F_i}\nabla f, \nabla_U F_i)\right)\\
&= \tfrac{2\lambda g(U,\nabla f)}{\|\nabla f\|^4}\alpha_i(\lambda - \alpha_i),
\end{aligned}
$$

$$
\begin{aligned}
&U\left(R(F_i, F_j, F_i, \nabla f)g(F_j, \nabla f)\right)\\
&\quad= \{(\nabla_U R)(F_i, F_j, F_i, \nabla f) + R(\nabla_U F_i, F_j, F_i, \nabla f) + R(F_i, \nabla_U F_j, F_i, \nabla f)\\
&\qquad + R(F_i, F_j, \nabla_U F_i, \nabla f) + R(F_i, F_j, F_i, \nabla_U \nabla f)\}g(F_j, \nabla f)\\
&\qquad + R(F_i, F_j, F_i, \nabla f)\left(g(\nabla_U F_j, \nabla f) + g(F_j, \nabla_U \nabla f)\right)\\
&\quad= \{-(\nabla_{F_i} R)(F_j, U, F_i, \nabla f) - (\nabla_{F_j} R)(U, F_i, F_i, \nabla f)\\
&\qquad + R(\nabla_U F_i, F_j, F_i, \nabla f) + R(F_i, \nabla_U F_j, F_i, \nabla f) + R(F_i, F_j, \nabla_U F_i, \nabla f)\\
&\qquad + R(F_i, F_j, F_i, \lambda U)\}g(F_j, \nabla f)\\
&\qquad + R(F_i, F_j, F_i, \nabla f)\left(g(\nabla_U F_j, \nabla f) + \lambda g(F_j, U)\right)\\
&\quad= \{R(\nabla_U F_i, F_j, F_i, \nabla f) + R(F_i, \nabla_U F_j, F_i, \nabla f)\\
&\qquad + R(F_i, F_j, \nabla_U F_i, \nabla f)\}g(F_j, \nabla f) + R(F_i, F_j, F_i, \nabla f)g(\nabla_U F_j, \nabla f) = 0\,.
\end{aligned}
$$

Consequently, along the slice S we have:

$$
\begin{aligned}
&U\left(R(F_i, F_j, F_i, V)g(F_j, V)\right) = U\|\nabla f\|^{-2}R(F_i, F_j, F_i, \nabla f)g(F_j, \nabla f)\\
&\qquad + \|\nabla f\|^{-2}U\left(R(F_i, F_j, F_i, \nabla f)g(F_j, \nabla f)\right) = 0,\\
&UR(F_i, F_j, F_i, F_j) = (\nabla_U R)(F_i, F_j, F_i, F_j) + 2R(\nabla_U F_i, F_j, F_i, F_j)\\
&\qquad + 2R(F_i, \nabla_U F_j, F_i, F_j)\\
&\quad= -(\nabla_{F_i} R)(F_j, U, F_i, F_j) - (\nabla_{F_j} R)(U, F_i, F_i, F_j)\\
&\qquad + 2R(\nabla_U F_i, F_j, F_i, F_j) + 2R(F_i, \nabla_U F_j, F_i, F_j)\\
&\quad= 2R(\nabla_U F_i, F_j, F_i, F_j) + 2R(F_i, \nabla_U F_j, F_i, F_j) = 0\,.
\end{aligned}
$$

Consequently, $0 = 2\lambda g(U, \nabla f)\|\nabla f\|^{-4}\alpha_i(\lambda - \alpha_i)$. Since λ and $g(U, \nabla f)$ are different from 0, either $\alpha_i = 0$ or $\alpha_i = \lambda$ for $i = 1, \ldots, m$. Since the level set S of f which was chosen was arbitrary, this is true on all of M. By Equation (11.1.b) we have $\mathcal{H}_f + \text{Ric} = \lambda\ \text{Id}$. The remaining conclusions of Assertion 3 are now immediate from the discussion above.

We now verify Assertion 4. Recall that (M, g) has a harmonic Weyl tensor if its Schouten tensor $S = \rho - \frac{\tau}{2(n+1)}g$ is Codazzi, i.e., $\nabla_X S_{YZ} = \nabla_Y S_{XZ}$ (see Besse [13]). If the Weyl tensor is harmonic then $(\nabla_X \rho)(Y, Z) - (\nabla_Y \rho)(X, Z) = 0$ since the scalar curvature is constant. Choose E_1, E_2 in Image$\{\mathcal{H}_f\}$ and F in Image{Ric}. We use Assertion 3 to see that

$$
0 = (\nabla_{E_1}\rho)(F, E_2) - (\nabla_F \rho)(E_1, E_2) = \rho(F, \nabla_{E_1} E_2) = \lambda g(F, \nabla_{E_1} E_2)\,.
$$

Choose E in Image$\{\mathcal{H}_f\}$ and F_1, F_2 in Image{Ric}. We show the two eigenspaces are parallel and that the soliton is rigid by computing:

$$\begin{aligned}
&(\nabla_{F_1}\rho)(E,F_2)-(\nabla_E\rho)(F_1,F_2)\\
&\quad=\rho(\nabla_{F_1}E,F_2)-E\rho(F_1,F_2)+\rho(\nabla_E F_1,F_2)+\rho(F_1,\nabla_E F_2)\\
&\quad=\lambda g(\nabla_{F_1}E,F_2)-\lambda E\,g(F_1,F_2)+\lambda g(\nabla_E F_1,F_2)+\lambda g(F_1,\nabla_E F_2)\\
&\quad=\lambda g(\nabla_{F_1}E,F_2)\,.
\end{aligned}$$

We complete the proof by verifying that Assertion 5 holds. We apply Theorem 11.16. If $\dim(\ker\{\mathrm{Ric}\})=2$, then $\mathcal{V}=\ker\{\mathrm{Ric}\}$. Since U is parallel, $\mathcal{H}_f(X)=\nabla_X\nabla f=\lambda X$ if X belongs to $\mathcal{V}$ and $\mathcal{H}_f(X)=\nabla_X\nabla f=0$ if $X\in\ker\{\mathcal{H}_f\}=\mathrm{Image}\{\mathrm{Ric}\}$. Consequently, the distribution $\mathcal{V}$ is parallel. Since the metric is not degenerate on $\mathcal{V}$, this implies that the manifold locally decomposes as a product $B\times F$ so that B is Ricci flat and, consequently, flat. On the other hand F is Einstein satisfying $\rho^F=\lambda g^F$. Therefore, the soliton is rigid. This completes the proof of Theorem 11.18. □

Theorem 11.18 leads to the following classification result in low dimensions.

Theorem 11.19 *Let (M,g,f) be a locally homogeneous Lorentzian non-steady gradient Ricci soliton of dimension $m\leq 4$. Then M is rigid.*

Proof. By Lemma 11.14, isotropic non-steady locally homogeneous gradient Ricci solitons are Einstein manifolds. We will therefore assume M is a non-isotropic non-steady locally homogeneous gradient Ricci soliton. If $m=3$, then Theorem 11.19 follows from the discussion above since $\dim(\ker\{\mathrm{Ric}\})=2$. We therefore assume $m=4$. The previous discussion shows we may assume $\dim(\ker\{\mathrm{Ric}\})=3$. We are going to use Theorem 11.18 to show that Image{Ric} is a non-null parallel distribution. We consider the adapted basis $\{U,\nabla f,E,F\}$ where $\{U,\nabla f,E\}$ is a basis of ker{Ric} and $F\cdot\mathbb{R}=\mathrm{Image}\{\mathrm{Ric}\}$. We show that the Weyl tensor is harmonic and (M,g,f) is rigid by examining the components of the curvature tensor which have ∇f as an argument:

$$\begin{aligned}
&R(E,\nabla f,\nabla f,E)=(\nabla_E\rho)(\nabla f,E)-(\nabla_{\nabla f}\rho)(E,E)=0,\\
&R(F,\nabla f,\nabla f,F)=(\nabla_F\rho)(\nabla f,F)-(\nabla_{\nabla f}\rho)(F,F)=0,\\
&R(F,\nabla f,E,\nabla f)=\rho(F,E)\|\nabla f\|^2=0,\\
&R(F,E,F,\nabla f)=\rho(\nabla f,E)=0,\quad R(E,F,E,\nabla f)=\rho(\nabla f,F)=0\,.
\end{aligned}$$

□

We have shown in proving Theorem 11.19 that if the factor N_0 of the decomposition given in Theorem 11.16 is of dimension $n_0\leq 4$ then the gradient Ricci soliton is rigid.

11.3.3 STEADY LOCALLY HOMOGENEOUS LORENTZIAN GRADIENT RICCI SOLITONS. The geometry of the level sets of the potential function plays an essential role in our analysis; the norm $\|\nabla f\|^2$ is important as this controls the nature of the metric on the level sets. The 2-dimensional case is trivial; one has Brozos-Vázquez, García-Río and Gavino-Fernández [17] and Chow et al. [47].

Theorem 11.20 *A steady locally homogeneous gradient Ricci soliton of dimension 2 either in the Riemannian or in the Lorentzian setting is flat.*

In higher dimensions, one has the following result.

Theorem 11.21 *Let (M, g, f) be a steady locally homogeneous gradient Lorentzian Ricci soliton. If $\|\nabla f\|^2 < 0$, then (M, g) splits locally as an isometric product $(\mathbb{R} \times N, -dt^2 + g_N)$ where (N, g_N) is a flat Riemannian manifold and f is orthogonal projection on $\mathbb{R}$.*

Proof. As before, we will use Lemma 11.14 throughout the section without further citation. Let (M, g, f) be a steady locally homogeneous Lorentzian gradient Ricci soliton. Then $\|\operatorname{Hess}_f\|^2 = 0$ and $\|\nabla f\|^2 = \mu$ is constant. In what follows we will consider the possibilities $\mu < 0$ and $\mu = 0$ separately. Assume that $\mu < 0$. As $\mathcal{H}_f(\nabla f) = 0$, we may restrict $\mathcal{H}_f$ to $\nabla f^\perp$. As $\nabla f^\perp$ inherits a positive definite metric and as $\|\operatorname{Hess}_f\|^2 = 0$, $\mathcal{H}_f = 0$. This shows that ∇f is a parallel vector field, and thus (M, g) is locally a product $(\mathbb{R} \times N, -dt^2 + g_N)$ where (N, g_N) is a locally homogeneous Riemannian manifold (see, for example, García-Río and Kupeli [70]). Additionally, (N, g_N) is a steady gradient Ricci soliton, and therefore Ricci flat. Spiro [126] showed that locally homogeneous Ricci flat Riemannian manifolds are locally isometric to Euclidean space. □

The cases when $\|\nabla f\|^2 \geq 0$ are less rigid in the steady setting. Several examples in the spacelike case $\|\nabla f\|^2 > 0$ are known (see, for example, Batat et al. [10] and Brozos-Vázquez, García-Río and Gavino-Fernández [17]). However, little more of a general nature is known about this case. In the isotropic case one has some restrictions on the Ricci operator; in particular, it must be nilpotent. Recall by Definition 10.21 that a tensor T is said to be *recurrent* if there is a smooth 1-form ω so that $\nabla_X T = \omega(X)T$.

Theorem 11.22 *Assume that (M, g, f) is a steady isotropic locally homogeneous gradient Lorentzian Ricci soliton. One of the following two possibilities pertains.*

1. $\mathcal{H}_f = -\operatorname{Ric}$ *has rank* 2 *and is* 3*-step nilpotent.*
2. $\mathcal{H}_f = -\operatorname{Ric}$ *has rank* 1 *and is* 2*-step nilpotent. In this case (M, g) is locally a strict Walker manifold. This means that:*
   (a) $\ker\{\mathcal{H}_f\} = \nabla f^\perp$ *and* $\operatorname{Image}\{\mathcal{H}_f\} = \nabla f$.
   (b) ∇f is a recurrent vector field and $\nabla f^\perp$ is an integrable totally geodesic distribution whose leafs are the level sets of f.
   (c) Let $P \in M$. At least one of the following possibilities holds near P.
      (i) There is a Killing vector field F so $\nabla\{F(f)\}$ is a null parallel vector field.
      (ii) There is a smooth function ψ defined near P so $\psi\ \nabla f$ is a null parallel vector field.

Proof. To prove Assertion 1, we assume that $\|\nabla f\|^2 = 0$ so ∇f is a null vector. Choose an orthonormal basis $\{E_1, \dots, E_{m+2}\}$ for the tangent space at a point so E_1 is timelike, so $\{E_2, \dots, E_{m+2}\}$ are spacelike, and so $\nabla f = c(E_1 + E_2)$ for some $c \neq 0$. We further normalize the basis so $\mathcal{H}_f E_1 \in \text{span}\{E_1, E_2, E_3\}$. Let $\mathcal{H}_f E_i = \mathcal{H}_i^j E_j$. Since $E_1 + E_2 \in \ker\{\mathcal{H}_f\}$, $\mathcal{H}_1^i + \mathcal{H}_2^i = 0$ for all i. Furthermore, $\mathcal{H}_1^i = \mathcal{H}_2^i = 0$ for $i \geq 4$ since $\mathcal{H}_f E_1 \in \text{span}\{E_1, E_2, E_3\}$. Finally, as $\mathcal{H}_f$ is self-adjoint, $\mathcal{H}_1^i = -\mathcal{H}_i^1$ for $2 \leq i$ and $\mathcal{H}_i^j = \mathcal{H}_j^i$ for $2 \leq i, j$. We summarize these relations:

$$\begin{aligned} &\mathcal{H}_1^i = -\mathcal{H}_i^1 \quad \text{for} \quad i \geq 2, \qquad \mathcal{H}_i^j = \mathcal{H}_j^i \quad \text{for} \quad 2 \leq i, j, \\ &\mathcal{H}_1^i = \mathcal{H}_2^i = 0 \quad \text{for} \quad i \geq 4, \quad \mathcal{H}_1^i + \mathcal{H}_2^i = 0 \quad \text{for all} \quad i\,. \end{aligned} \tag{11.3.b}$$

Since $\mathcal{H}_f = \mathcal{H}_i^j E^i \otimes E_j$ and $\|\operatorname{Hess}_f\|^2 = \lambda((m+2)\lambda - \tau) = 0$, we have

$$0 = \|\operatorname{Hess}_f\|^2 = \|\mathcal{H}_f\|^2 = (\mathcal{H}_1^1)^2 - 2\sum_{i\geq 2}(\mathcal{H}_i^1)^2 + \sum_{2\leq j,k}(\mathcal{H}_j^k)^2\,. \tag{11.3.c}$$

The relations of Equation (11.3.b) then permit us to rewrite Equation (11.3.c) in the form:

$$0 = \sum_{3\leq j,k}(\mathcal{H}_j^k)^2\,.$$

This implies that $\mathcal{H}_j^k = 0$ for $3 \leq j, k$. Thus, by Equation (11.3.b), $\mathcal{H}_f E_i = 0$ for $i \geq 4$. Consequently, the relevant portion of the matrix $\mathcal{H}$ is given by:

$$\mathcal{H} = \begin{pmatrix} \mathcal{H}_1^1 & \mathcal{H}_2^1 & \mathcal{H}_3^1 \\ \mathcal{H}_1^2 & \mathcal{H}_2^2 & \mathcal{H}_3^2 \\ \mathcal{H}_1^3 & \mathcal{H}_2^3 & \mathcal{H}_3^3 \end{pmatrix} = \begin{pmatrix} \mathcal{H}_1^1 & -\mathcal{H}_1^1 & \mathcal{H}_3^1 \\ \mathcal{H}_1^1 & -\mathcal{H}_1^1 & \mathcal{H}_3^1 \\ -\mathcal{H}_3^1 & \mathcal{H}_3^1 & 0 \end{pmatrix}.$$

We compute

$$\mathcal{H}^2 = (\mathcal{H}_1^3)^2 \begin{pmatrix} -1 & 1 & 0 \\ -1 & 1 & 0 \\ 0 & 0 & 0 \end{pmatrix} \quad \text{and} \quad \mathcal{H}^3 = 0\,.$$

This shows that $\mathcal{H}$ is either 2-step or 3-step nilpotent which proves Assertion 1.

If $\mathcal{H}_f$ be 2-step nilpotent, then ∇f belongs to Image$\{\mathcal{H}_f\}$. Since $\mathcal{H}_f$ has rank 1, Image$\{\mathcal{H}_f\} = \nabla f \cdot \mathbb{R}$. We use the Fredholm alternative and the fact that $\mathcal{H}_f$ is self-adjoint to establish Assertion 2-a using the following equivalencies:

$$\begin{aligned} &\mathcal{H}_f Z = 0 && \Leftrightarrow \quad g(\mathcal{H}_f Z, Y) = 0 \quad \forall\, Y \\ \Leftrightarrow \quad &g(Z, \mathcal{H}_f Y) = 0 \quad \forall\, Y && \Leftrightarrow \quad Z \perp \text{range}\{\mathcal{H}_f\} \qquad \Leftrightarrow \quad Z \perp \nabla f\,. \end{aligned}$$

Choose a vector field U so $g(U, \nabla f) = 1$. Since range$\{\mathcal{H}_f\} = \nabla f$ and since $g(U, \nabla f) = 1$, we may establish that ∇f is recurrent by computing:

$$\nabla_X(\nabla f) = \mathcal{H}_f(X) = \theta(X) \cdot \nabla f \quad \text{where} \quad \theta(X) = g(U, \mathcal{H}_f(X))\,. \tag{11.3.d}$$

Let X and Y be smooth vector fields which are perpendicular to ∇f. We show that $[X,Y]$ is orthogonal to $\nabla f^{\perp}$ and thereby establish that $\nabla f^{\perp}$ is an integrable distribution by computing:

$$\begin{aligned} g([X,Y],\nabla f) &= g(\nabla_X Y - \nabla_Y X, \nabla f) \\ &= Xg(Y,\nabla f) - g(Y,\nabla_X\nabla f) - Yg(X,\nabla f) + g(X,\nabla_Y\nabla f) \\ &= X\{0\} - \operatorname{Hess}_f(Y,X) - Y\{0\} + \operatorname{Hess}_f(X,Y) = 0\,. \end{aligned}$$

Let $\gamma(t)$ be a geodesic with $\dot\gamma(0) \perp \nabla f$. We compute

$$\partial_t g(\dot\gamma,\nabla f) = g(\ddot\gamma,\nabla f) + g(\dot\gamma,\nabla_{\partial_t}\nabla f) = \theta(\partial_t)g(\dot\gamma,\nabla f)\,.$$

Since $g(\dot\gamma,\nabla f)(0)=0$, the fundamental theorem of ODEs implies that $g(\dot\gamma,\nabla f)$ vanishes identically. Consequently, $\dot\gamma \in \nabla f^{\perp}$. Since $g(\dot\gamma,\nabla f) = \partial_t f$, the geodesic lies entirely in the level set of f. Assertion 2-b follows.

We proceed by induction on the dimension to establish Assertion 2-c. Fix a point $P \in M$. Let $\mathcal{V} := \operatorname{span}\{U,\nabla f\}$. The metric on $\mathcal{V}$ is non-degenerate and contains a null vector. This shows that $\mathcal{V}$ has Lorentzian signature. We can choose complementary Killing vector fields $\{F_1,\dots,F_m\}$ so $\{U,\nabla f,F_1,\dots,F_m\}$ is a local frame field near P and so that

$$g(U,F_i)|_P = g(\nabla f,F_i)|_P = 0\,. \tag{11.3.e}$$

Consequently, $\operatorname{span}\{F_1,\dots,F_m\}$ is spacelike near P. Let $\xi_i := \nabla\{F_i(f)\}$; these are parallel vector fields by Lemma 11.14. Let $\mathcal{B} := \operatorname{span}\{\xi_1,\dots,\xi_m\}$. Since the ξ_i are parallel,

$$r(x) := \operatorname{Rank}(\mathcal{B}(x))$$

is locally constant. Suppose $r>0$. By reordering the collection $\{F_1,\dots,F_m\}$ if necessary, we may assume that $\{\xi_1,\dots,\xi_r\}$ is a local frame field for $\mathcal{B}$. Let $\epsilon_{ij} := g(\xi_i,\xi_j)$ describe the induced metric on $\mathcal{B}$. Again, we use the fact that the elements ξ_i of the frame field are parallel; this implies that the ϵ_{ij} are constant. We can diagonalize ϵ or equivalently renormalize the choice of the Killing vector fields F_i to assume that ϵ is in fact diagonal. If $\det(\epsilon)=0$, then ξ_i is a parallel null vector field for some i and Assertion 2-c-i holds. Thus, we may assume that the inner product restricted to $\mathcal{B}$ is non-degenerate. We may use Theorem 11.13 to decompose, at least locally, $M = N^{2+m-r}\times\mathbb{R}^r_\nu$. If the metric on N is Riemannian, we may apply Theorem 11.13 to see that the soliton is trivial. Thus, N is Lorentzian. If $\dim(N)=2$, then Theorem 11.20 shows N is flat and $\mathcal{H}_f=0$ which is false. This shows $\dim(N)\ge 3$ and we may use our induction hypothesis on N. Thus, we may assume without loss of generality that $r=0$ so $\mathcal{B}=\{0\}$ and assume henceforth that $\nabla\{F_i(f)\}=0$ for all i. This shows that $\kappa_i := F_i(f)$ is constant for all i. By Equation (11.3.e), $\kappa_i = F_i(f)|_P = g(F_i,\nabla f)|_P = 0$. This shows that $g(F_i,\nabla f)$ vanishes identically and we have

$$F_i \in \ker\{\mathcal{H}_f\} = \ker\{\operatorname{Ric}\} = \nabla f^{\perp}\,. \tag{11.3.f}$$

We apply Equation (11.3.d) and Equation (11.3.f) to see that

$$\begin{aligned} &\nabla_{\nabla f}\nabla f = \mathcal{H}_f(\nabla f) = 0, \quad \nabla_{F_i}\nabla f = \mathcal{H}_f(F_i) = 0 \quad \text{for all} \quad i, \\ &\nabla_U \nabla f = \mathcal{H}_f(U) = \Xi \nabla f \quad \text{where} \quad \Xi := g(\mathcal{H}_f(U), U) = -\rho(U,U). \end{aligned} \tag{11.3.g}$$

We use Equation (11.3.g) to see $\nabla_Y \nabla f = 0$ if $Y \perp \nabla f$. Thus, the only undetermined covariant derivative $\nabla_U \nabla f$. Let $\Psi := \psi \cdot \nabla f$. This is a null vector field and Ψ will be parallel if and only if ψ satisfies the relations:

$$Y(\psi) = 0 \quad \text{if} \quad Y \perp \nabla f \quad \text{and} \quad U(\psi) + \psi \Xi = 0. \tag{11.3.h}$$

Since F_i is a Killing vector field, $\nabla_{F_i}\rho = 0$. Since $F_i \in \ker\{\text{Ric}\}$, $\rho(F_i, \cdot)$ vanishes identically. Consequently, we may use Lemma 11.14 to see that:

$$\begin{aligned} R(F_i, U, \nabla f, F_j) &= (\nabla_{F_i}\rho)(U, F_j) - (\nabla_U \rho)(F_i, F_j) \\ &= -U\rho(F_i, F_i) + \rho(\nabla_U F_i, F_j) + \rho(\nabla_U F_j, F_i) = 0. \end{aligned} \tag{11.3.i}$$

Let $g_{ij} = g(F_i, F_j)$. Since $U \in \ker\{\text{Ric}\}$ and since $\{U, \nabla f\}$ are a hyperbolic pair, we may use Equation (11.3.i) to see that:

$$\begin{aligned} 0 &= \rho(U, \nabla f)|_P = R(U, \nabla f, \nabla f, U)|_P + \sum_{i,j=1}^{m} g^{ij} R(U, F_i, \nabla f, F_j)|_P \\ &= R(U, \nabla f, \nabla f, U)|_P. \end{aligned}$$

Since P was arbitrary and the only condition on U was that $g(U, \nabla f) = 1$, this holds for arbitrary P and we have $R(U, \nabla f, \nabla f, U) = 0$ if $g(U, \nabla f) = 1$. If X is a Killing vector field and if Y and Z are arbitrary, we have (see, for example, Kostant [87] or Nomizu [108]) that $\mathcal{R}(X, Y)Z = -\nabla_Y \nabla_Z X + \nabla_{\nabla_Y Z} X$. Let Ξ be as defined in Equation (11.3.g). We use Equation (11.3.f) to see

$$g(\nabla_U F_i, \nabla f) = U\, g(F_i, \nabla f) - g(F_i, \nabla_U \nabla f) = -g(F_i, \Xi \nabla f) = 0.$$

Since the F_i are Killing vector fields, since $g(F_i, \nabla f) = 0$, and since ∇f is recurrent,

$$\begin{aligned} R(F_i, U, U, \nabla f) &= -g(\nabla_U \nabla_U F_i, \nabla f) + g(\nabla_{\nabla_U U} F_i, \nabla f) \\ &= -U\, g(\nabla_U F_i, \nabla f) + g(\nabla_U F_i, \nabla_U \nabla f) \\ &\quad +(\nabla_U U) g(F_i, \nabla f) - g(F_i, \nabla_{\nabla_U U}\{\nabla f\}) \\ &= -U\{U\, g(F_i, \nabla f) - g(F_i, \nabla_U \nabla f)\} + g(\nabla_U F_i, \Xi \nabla f) \\ &= U\, g(F_i, \Xi \nabla f) + \Xi g(\nabla_U F_i, \nabla f) = 0. \end{aligned}$$

By Lemma 11.14, if $\{X, Y, Z\}$ are vector fields on a gradient Ricci soliton, then

$$R(X, Y, \nabla f, Z) = (\nabla_X \rho)(Y, Z) - (\nabla_Y \rho)(X, Z).$$

Consequently, we have that

$$0 = R(U,\nabla f,\nabla f,U) = (\nabla_U\rho)(\nabla f,U) - (\nabla_{\nabla f}\rho)(U,U),$$
$$0 = R(F_i,U,\nabla f,U) = (\nabla_{F_i}\rho)(U,U) - (\nabla_U\rho)(F_i,U).$$

By Equation (11.3.g), $\Xi = -\rho(U,U)$. Thus, we may compute:

$$\begin{aligned}
-\nabla f(\Xi) &= \nabla f\rho(U,U) = (\nabla_{\nabla f}\rho)(U,U) + 2\,\rho(\nabla_{\nabla f}U,U)\\
&= (\nabla_U\rho)(\nabla f,U) - 2\,g(\nabla_{\nabla f}U,\Xi\nabla f)\\
&= U\,\rho(\nabla f,U) - \rho(\nabla_U\nabla f,U) - \rho(\nabla f,\nabla_U U)\\
&\quad -2\Xi(\nabla f g(U,\nabla f) - g(U,\nabla_{\nabla f}\nabla f)) = 0, \quad \text{and}\\
-F_i(\Xi) &= F_i\rho(U,U) = (\nabla_{F_i}\rho)(U,U) + 2\,\rho(\nabla_{F_i}U,U)\\
&= (\nabla_U\rho)(F_i,U) - 2\,g(\nabla_{F_i}U,\Xi\nabla f)\\
&= U\,\rho(F_i,U) - \rho(\nabla_U F_i,U) - \rho(F_i,\nabla_U U)\\
&\quad -2\Xi(F_i g(U,\nabla f) - g(U,\nabla_{F_i}\nabla f))\\
&= g(\nabla_U F_i,\Xi\nabla f) = \Xi U g(F_i,\nabla f) - \Xi g(F_i,\Xi\nabla f) = 0\,.
\end{aligned}$$

This shows that $X(\Xi) = 0$ if $X \in \nabla f^{\perp}$. Since the distribution $\nabla f^{\perp}$ is integrable, the Frobenius theorem means we can introduce local coordinates $(u,x^2,\dots,x^{m+2})$ so that $U = \partial_u$ and $\nabla f^{\perp} = \operatorname{span}\{\partial_{x^2},\dots,\partial_{x^{m+2}}\}$. Thus, Equation (11.3.h) becomes an ordinary differential equation which can be solved. This completes the proof of Theorem 11.22. □

11.3.4 SYMMETRIC LORENTZIAN GRADIENT RICCI SOLITONS. There are stronger results which are available if (M,g) is *locally symmetric* or, equivalently, if $\nabla R = 0$. We generalize Definition 10.19 to the higher-dimensional setting as follows.

Definition 11.23 We say that (N,g_N) is a *Cahen–Wallach space* if there are coordinates $(t,y,x^1,\dots,x^m)$ and constants $0 \neq \kappa_i \in \mathbb{R}$ so:

$$g = 2\,dt \circ dy + (\kappa_1(x^1)^2 + \dots + \kappa_m(x^m)^2)dy^2 + (dx^1)^2 + \dots + (dx^m)^2.$$

Theorem 11.24

1. *Let (M,g) be a Lorentzian locally symmetric space.*
   (a) If (M,g) is irreducible, then (M,g) has constant sectional curvature.
   (b) If (M,g) is indecomposable but reducible, then (M,g) is a Cahen–Wallach space.
2. *If (M,g,f) is a Cahen–Wallach gradient Ricci soliton, then (M,g,f) is steady, $f = a_0 + a_1 y + \frac{1}{4}\sum_i \kappa_i y^2$, and $\nabla f = (a_1 + \frac{1}{2}\sum_i \kappa_i y)\partial_t$ is null.*

Proof. We refer to Cahen et al. [28] or Cahen and Wallach [29] for the proof of Assertion 1. We follow the discussion in Batat et al. [10] to prove Assertion 2. Let $\kappa := \kappa_1 + \cdots + \kappa_m$. The Levi–Civita connection of the metric given in Definition 11.23 is determined by the non-zero Christoffel symbols:

$$\nabla_{\partial_y}\partial_y = -\kappa_1 x^1 \partial_{x^1} + \cdots + \kappa_m x^m \partial_{x^m} \quad \text{and} \quad \nabla_{\partial_y}\partial_{x^i} = \nabla_{\partial_{x^i}}\partial_y = \kappa_i x^i \partial_v\,.$$

The only (possibly) non-zero entries in the curvature tensor are $R(\partial_y, \partial_{x^i}, \partial_y, \partial_{x^i}) = -\kappa_i$. Consequently, the only (possibly) non-zero entry in the Ricci tensor is $\rho(\partial_y, \partial_y) = -\kappa$. If $\kappa \neq 0$, then $\mathrm{Ric}\,(\partial_y) = -\kappa\partial_t$ and $\mathrm{Ric}\,(\partial_t) = 0$. Consequently, the Ricci tensor is 2-step nilpotent. Furthermore, f defines a gradient Ricci soliton if and only if

$$f(t, y, x^1, \ldots, x^m) = f(y) \quad \text{where} \quad f(y) = a_0 + a_1 y + \frac{1}{4}\kappa y^2\,.$$

We have $\lambda = 0$ in this instance. Note that $df = (a_1 + \frac{1}{2}\kappa y)dy$. Thus, $\nabla f = (a_1 + \frac{1}{2}\kappa y)\partial_t$ is a null parallel vector field. □

Theorem 11.24 will play a crucial role in the proof of the following result.

Theorem 11.25 *Let (M, g, f) be a locally symmetric Lorentzian gradient Ricci soliton. Then (M, g) splits locally as a product $M = N \times \mathbb{R}^k$ where one of the following possibilities holds.*

1. *If (M, g, f) is not steady, then (N, g_N) is an Einstein manifold and the soliton is rigid.*
2. *If (M, g, f) is steady, then (N, g_N, f_N) is locally isometric to a Cahen–Wallach space.*

Proof. Let (M, g) be a locally symmetric Lorentzian manifold. If (M, g, f) is a non-steady gradient Ricci soliton, then by Theorem 11.16, M splits, at least locally, as a product in the form $M = N_0 \times N_1 \times \mathbb{R}^k$ where (N_0, g_0) is indecomposable but reducible and (N_1, g_1) is an Einstein manifold. If N_0 does not appear in the decomposition, then the soliton is rigid. Otherwise, (N_0, g_0) is an indecomposable but not irreducible Lorentzian symmetric space. Therefore, it is a Cahen–Wallach space (see Cahen and Wallach [29] or Berard Bergery and Ikemakhen [11]). Theorem 11.24 rules out this latter possibility since if (N, g_N, f_N) is a Cahen–Wallach gradient Ricci soliton, then it is steady.

Next, suppose that (M, g, f) is a locally symmetric Lorentzian steady gradient Ricci soliton. We can use the de Rham–Wu decomposition of the manifold to split (M, g) locally as a product $M = N \times M_1 \times \cdots \times M_\ell \times \mathbb{R}^k_\nu$ where (N, g_N) is a Cahen–Wallach space, where the M_i are irreducible symmetric spaces, and where $\mathbb{R}^k_\nu$ is either Euclidean or Minkowski space. Since irreducible symmetric spaces are Einstein manifolds, the induced soliton is either trivial or the scalar curvature vanishes, which implies that M_i is Ricci flat. If M_i is Riemannian, then it is flat since Ricci flat locally symmetric spaces are flat in the Riemannian setting (see Besse [13] or Helgason [80]). Moreover, if M_i is Lorentzian, then it is flat since irreducible Lorentzian locally symmetric spaces are of constant sectional curvature (see Cahen et al. [28]). Hence, if the gradient Ricci soliton is steady, then the decomposition above reduces to $M = N \times \mathbb{R}^k$ where (N, g_N) is a Cahen–Wallach space. Theorem 11.25 now follows. □

11.3.5 3-DIMENSIONAL LOCALLY HOMOGENEOUS GRADIENT RICCI SOLITONS. Let (M, g) be a Lorentzian manifold of dimension 3. We suppose first that (M, g) is strict Walker, i.e., admits a null parallel vector field. We may then (see, for example, Brozos-Vázquez et al. [21]) find local adapted coordinates (t, x, y) so that the metric has the form which was given Definition 10.19:

$$g_\phi = 2dt \circ dy + dx^2 + \phi(x, y)dy^2\,. \tag{11.3.j}$$

The following is of independent interest; we drop for the moment the assumption that the metric is locally homogeneous and focus on Walker geometry.

Theorem 11.26 *Let $\mathcal{M}_\phi := (M, g_\phi)$ be a non-flat 3-dimensional Lorentzian strict Walker manifold. Then (M, g_ϕ, f) is a gradient Ricci soliton if and only if there exist a cover of M by coordinate systems where the metric has the form given in Equation (11.3.j) where one of the following occurs.*

1. *$\phi(x, y) = \frac{1}{\alpha^2}\, a(y)\, e^{\alpha x} + x\, b(y) + c(y)$ and $f(x, y) = x\,\alpha + \gamma(y)$ for $\gamma''(y) = -\frac{1}{2}\,\alpha b(y)$ and $\alpha \in \mathbb{R}$. In this setting, $\nabla f = \alpha\partial_x + \gamma'(y)\partial_t$ is spacelike.*
2. *$\phi(x, y) = x^2\, a(y) + x\, b(y) + c(y)$ and $f(x, y) = \gamma(y)$ where $\gamma''(y) = \frac{1}{4}\, a(y)$. In this setting, $\nabla f = \gamma'\partial_t$ is null.*

Moreover, in both cases the Ricci soliton is steady.

Proof. Let (M, g) be a 3-dimensional Lorentzian strict Walker metric. Choose local coordinates so the metric is given by Equation (11.3.j). Let $f(t, x, y)$ be a smooth real-valued function. To simplify the notation, set $f_t = \frac{\partial f}{\partial t}$, $f_{tx} = \frac{\partial^2 f}{\partial t \partial x}$, and so forth. One sees that the soliton equation $\mathrm{Hess}_f + \rho = \lambda g$ is equivalent to the following relations:

$$\begin{aligned} 0 &= f_{tt} = f_{tx}, & 0 &= f_{xx} - \lambda = f_{ty} - \lambda, \\ 0 &= 2f_{xy} - \phi_x f_t, & 0 &= 2\lambda\,\phi + \phi_{xx} - 2f_{yy} - \phi_x f_x + \phi_y f_t\,. \end{aligned} \tag{11.3.k}$$

This implies that $f(t, x, y) = t(\lambda\, y + \kappa) + \frac{1}{2}\,\lambda\, x^2 + \alpha(y)\, x + \gamma(y)$ for $\kappa \in \mathbb{R}$. Hence, the relations of Equation (11.3.k) simplify to become

$$0 = 2\alpha'(y) - (\lambda\, y + \kappa)\,\phi_x, \tag{11.3.l}$$

$$0 = 2\lambda\,\phi - 2\,\gamma''(y) - 2\,x\,\alpha''(y) + (\lambda\, y + \kappa)\,\phi_y - (\lambda\, x + \alpha(y))\,\phi_x + \phi_{xx}\,. \tag{11.3.m}$$

We differentiate Equation (11.3.l) with respect to x to conclude:

$$0 = (\lambda\, y + \kappa)\,\phi_{xx}\,. \tag{11.3.n}$$

Since the Ricci operator is given by

$$\text{Ric} = \begin{pmatrix} 0 & 0 & -\frac{1}{2}\phi_{xx} \\ 0 & 0 & 0 \\ 0 & 0 & 0 \end{pmatrix},$$

the metric is flat if and only if $\phi_{xx} = 0$. Since we have assumed that the Walker metric is not flat, we may use Equation (11.3.n) to see that $\lambda = \kappa = 0$ and to conclude that the gradient Ricci soliton is steady. Consequently, Equation (11.3.l) and Equation (11.3.n) imply that $f(t,x,y) = \alpha\, x + \gamma(y)$. Therefore, Equation (11.3.m) shows $2\gamma''(y) + \alpha\,\phi_x - \phi_{xx} = 0$. We differentiate with respect to x to see $\alpha\,\phi_{xx} = \phi_{xxx}$. This gives rise to two different cases.

Case I. Suppose that $\alpha \neq 0$. We then have $\phi(x,y) = \frac{1}{\alpha^2}\, a(y)\, e^{\alpha x} + x\, b(y) + c(y)$. Furthermore, the potential function of the soliton is $f(t,x,y) = \alpha\, x + \gamma(y)$ where $\gamma''(y) = -\frac{1}{2}\,\alpha\, b(y)$. We then have that $\nabla f = \gamma'(y)\,\partial_t + \alpha\,\partial_x$ is spacelike and Assertion 1 follows.

Case II. Suppose that $\alpha = 0$. We then have $\phi(x,y) = x^2\, a(y) + x\, b(y) + c(y)$ and the potential function of the soliton is given by $f(t,x,y) = \gamma(y)$ where $\gamma''(y) = \frac{1}{4}\, a(y)$. In this case $\nabla f = \gamma'(y)\,\partial_t$ is a null and recurrent vector field. Assertion 2 follows. □

We introduce some additional notation.

Definition 11.27 Let $g = g_\phi = 2dtdy + dx^2 + \phi(x,y)dy^2$ define a 3-dimensional strictly Walker manifold $\mathcal{M}_\phi$.

1. Let $\mathcal{N}_b$ be defined by taking $\phi(x,y) = b^{-2}e^{bx}$ for $0 \neq b \in \mathbb{R}$.
2. Let $\mathcal{P}_c$ be defined by taking $\phi(x,y) = \frac{1}{2}x^2\alpha(y)$ where $\alpha_y(y) = c\alpha^{3/2}(y)$ and $\alpha(y) > 0$.
3. Let $\mathcal{CW}_\pm$ be the Cahen–Wallach space defined by taking $\phi(x,y) = \pm x^2$.

In Section 10.4 (see Theorem 10.26), we showed that $\mathcal{M}_\phi$ is locally homogeneous if and only if we could choose local coordinates so that the function ϕ had one of the three forms given above in Definition 11.27. We can now state our classification result.

Theorem 11.28 *Let $\mathcal{M} = (M, g, f)$ be a Lorentzian locally homogeneous gradient Ricci soliton of dimension 3. If $\mathcal{M}$ is non-trivial, then either $\mathcal{M}$ is rigid or $\mathcal{M}$ is locally isometric to either $\mathcal{CW}_\pm$, $\mathcal{P}_c$ or $\mathcal{N}_b$ as defined above and the soliton is steady.*

1. *∇f is null if $(M,g) = \mathcal{P}_c$ or if $(M,g) = \mathcal{CW}_\pm$.*
2. *∇f is spacelike if $(M,g) = \mathcal{N}_b$.*

Proof. We distinguish the following cases.

Case I. Suppose that $\mathcal{M}$ is non-steady. By Theorem 11.19, $\mathcal{M}$ is rigid.

Case II. Suppose that $\mathcal{M}$ is steady. By Lemma 11.14, the potential function is a solution of the Eikonal equation $\|\nabla f\|^2 = \mu$. We distinguish three subcases.

Case II-a. $\mathcal{M}$ is steady and $\mu < 0$. Then $\mathcal{M}$ splits locally as a product so $\mathcal{M}$ is rigid.

Case II-b. $\mathcal{M}$ is steady and $\mu = 0$. The Ricci operator is either 2-step or 3-step nilpotent. It follows from work of Calvaruso and Kowalski [32] that there do not exist locally homogeneous 3-dimensional manifolds with 3-step nilpotent Ricci operator. Consequently, the Ricci operator is 2-step nilpotent and $\mathcal{M}$ admits a locally defined parallel null vector field. This shows that $\mathcal{M}$ is locally a strict Walker manifold. Therefore, the underlying geometry of (M, g) is given by Theorem 10.26; the function f is now determined by Theorem 11.26.

Case II-c. $\mathcal{M}$ is steady and $\mu > 0$. Since the scalar curvature is constant, the Ricci operator satisfies $\mathrm{Ric}(\nabla f) = 0$, which shows that either f is constant, or otherwise the Ricci operator has a zero eigenvalue. We now consider the different possibilities for the kernel of the Ricci operator.

Assume $\dim(\ker\{\mathrm{Ric}\}) = 1$. It follows from Calvaruso [30] that (M, g) is either a symmetric space or a Lie group. If $\mathcal{M}$ is symmetric, then it is one of the following: a manifold of constant sectional curvature, a product $\mathbb{R} \times N$ where (N, g_N) is of constant curvature, or a 3-dimensional Cahen–Wallach space. Hence, in all the cases, any gradient Ricci soliton is trivial, rigid or the underlying manifold admits a null parallel vector field (and we have already examined that case). Now we concentrate on Lie groups. Since the eigenspaces of the Ricci operator are left-invariant, since ∇f has constant norm $\mu > 0$, and since $\dim(\ker\{\mathrm{Ric}\}) = 1$, we conclude ∇f is a left-invariant vector field. Left-invariant Ricci solitons on 3-dimensional Lorentzian Lie groups were considered in Brozos-Vázquez et al. [15]. They showed that they exist if and only if the Ricci operator has exactly one-single eigenvalue, which must be zero since $\mathrm{Ric}(\nabla f) = 0$. This shows that the Ricci operator is 3-step nilpotent, but that is not possible due to the analysis carried out in Calvaruso and Kowalski [32].

Finally, assume $\dim(\ker\{\mathrm{Ric}\}) = 2$. In this case the Ricci operator is either diagonalizable or 2-step nilpotent. The latter implies that the manifold admits locally a null parallel vector field (see Calviño-Louzao et al. [35]), and, again, this case has been treated. If the Ricci operator is diagonalizable, then $\| \mathrm{Ric} \|^2 = \pm\tau^2 = \| \mathrm{Hess}_f \|^2$ and Assertion 3 of Lemma 11.14 shows that $\tau = 0$, from where it follows that (M, g) is flat and the soliton is trivial. This completes the proof of Theorem 11.28. □

Work of di Cerbo [41] shows that 3-dimensional Lie groups do not admit left-invariant Riemannian Ricci solitons. The Lorentzian case is much richer, allowing the existence of expanding, steady and shrinking left-invariant Ricci solitons. The following summarizes the classification of left-invariant Lorentzian Ricci solitons in dimension 3 given in Brozos-Vázquez et al. [15] where we adopt the notation of that paper.

Theorem 11.29 *Let (G, g) be a 3-dimensional Lie group equipped with a left-invariant Lorentzian metric admitting a left-invariant Ricci soliton and let $\{e_1, e_2, e_3\}$ be an orthonormal basis of the corresponding Lie algebra of signature $(+ + -)$.*

1. *G is a unimodular Lie group corresponding to one of the following Lie algebras:*

(a) $[e_1, e_2] = \frac{1}{2}e_2 - (\beta - \frac{1}{2})e_3$, $[e_1, e_3] = -(\beta + \frac{1}{2})e_2 - \frac{1}{2}e_3$, $[e_2, e_3] = \alpha e_1$. *One has*

(i) $\alpha = 0$ *and* $G = E(1,1)$. *Ricci solitons are steady and the left-invariant vector field is given by* $X = -\beta e_1$.

(ii) $\alpha = \beta \neq 0$ *and* $G = SL(2,\mathbb{R})$. *Ricci solitons are expanding and there exists a one-parameter family of left-invariant Ricci solitons given by setting, for any* $t \in \mathbb{R}$, $X = -\frac{1}{2}\beta e_1 + te_2 + te_3$.

(b) $[e_1, e_2] = -\frac{1}{\sqrt{2}}e_1 - \alpha e_3$, $[e_1, e_3] = -\frac{1}{\sqrt{2}}e_1 - \alpha e_2$, $[e_2, e_3] = \alpha e_1 + \frac{1}{\sqrt{2}}e_2 - \frac{1}{\sqrt{2}}e_3$. *If* $\alpha = 0$ *then* $G = E(1,1)$. *If* $\alpha \neq 0$, *then either* $G = (1,2)$ *or* $G = SL(2,\mathbb{R})$. *In this instance, the Ricci solitons are expanding and the left-invariant vector field is given by setting* $X = \alpha e_1 - \frac{1}{\sqrt{2}}e_2 + \frac{1}{\sqrt{2}}e_3$.

2. *G is a non-unimodular Lie group G corresponding to one of the following the Lie algebras:* $[e_1, e_2] = -\frac{1}{\sqrt{2}}(\alpha e_1 + \frac{1}{\sqrt{2}}\beta(e_2 + e_3))$, $[e_1, e_3] = \frac{1}{\sqrt{2}}(\alpha e_1 + \frac{1}{\sqrt{2}}\beta(e_2 + e_3))$, $[e_2, e_3] = \frac{1}{\sqrt{2}}\delta(e_2 + e_3)$. *Ricci solitons are steady and the left-invariant vector field is given by* $X = \frac{\alpha^2 - \alpha\delta}{2\delta\sqrt{2}}(e_2 + e_3)$. *If* $\alpha = \frac{1}{2}\delta$, *the left-invariant Ricci solitons can be expanding, steady or shrinking, depending on the value of* λ *and are given by*

$$X = -\frac{2\beta\lambda}{\delta^2}e_1 - \frac{\delta^4 + 8(\delta^2 - 2\beta^2)\lambda}{8\delta^3\sqrt{2}}e_2 - \frac{\delta^4 - 8(\delta^2 + 2\beta^2)\lambda}{8\delta^3\sqrt{2}}e_3 .$$

The classification of homogeneous Ricci almost solitons is more complicated in the Lorentzian setting than it is in the Riemannian setting. For example, Haji-Badali [79] has shown that 3-dimensional homogeneous Walker manifolds $\mathcal{M}_f$ with $f = b^{-2}e^{by}$ (cf. Theorem 10.26) admit Ricci almost solitons, although they are not locally symmetric.

11.4 RIEMANNIAN LOCALLY CONFORMALLY FLAT GRADIENT RICCI SOLITONS

The nature of the Ricci tensor (non-negative or non-positive) plays an important role in the classification of locally conformally flat manifolds; see, for example, the discussion in Zhu [137]. In the present instant, the type of Ricci soliton (expanding, steady, shrinking) plays a similar role in the analysis. As the expanding situation is quite open, we will focus our attention on the shrinking and steady cases.

We follow Fernández-López and García-Río [64] to show that a locally conformally flat gradient Ricci soliton (shrinking, steady or expanding, and not necessarily complete) is locally a warped product. Note that a Riemannian manifold of dimension $m \geq 4$ is locally conformally flat if and only if its Weyl tensor vanishes. In dimension $m = 3$ the Weyl tensor is always zero, and the manifold is locally conformally flat if and only if the Schouten tensor is a Codazzi tensor (which is a consequence of local conformal flatness in higher dimensions).

Lemma 11.30 Let $\mathcal{M} = (M, g, f)$ be an m-dimensional locally conformally flat non-trivial gradient Ricci soliton. Then wherever $\nabla f \neq 0$, $\mathcal{M}$ is locally isometric to a warped product $(M, g) = ((a, b) \times N, dt^2 + \psi(t)^2 g_N)$ of an interval with a Riemannian manifold (N, g_N) of constant sectional curvature.

Proof. Since (M, g) is locally conformally flat, the Weyl tensor W vanishes. We have that the Schouten tensor $\mathfrak{S} = \rho - \frac{\tau}{2(m-1)} g$ is a Codazzi tensor, or, equivalently,

$$(\nabla_X \mathfrak{S})(Y, Z) = (\nabla_Y \mathfrak{S})(X, Z)$$

for all vector fields X, Y, Z. (If $m = 3$, this condition is equivalently to local conformal flatness). We may then compute that:

$$\begin{aligned}&(\nabla_X \rho)(Y, Z) - \tfrac{1}{2(m-1)} X(\tau) g(Y, Z) - \tfrac{1}{2(m-1)} \tau(\nabla_X g)(Y, Z)\\ &\quad = (\nabla_Y \rho)(X, Z) - \tfrac{1}{2(m-1)} Y(\tau) g(X, Z) - \tfrac{1}{2(m-1)} \tau(\nabla_Y g)(X, Z)\,.\end{aligned}$$

By Lemma 11.5,

$$R(X, Y, Z, \nabla f) = \tfrac{1}{m-1} \rho(X, \nabla f) g(Y, Z) - \tfrac{1}{m-1} \rho(Y, \nabla f) g(X, Z)\,. \tag{11.4.a}$$

We set $Z = \nabla f$ and $Y \perp \nabla f$ in Equation (11.4.a) to see that ∇f is an eigenvector of both the Ricci operator and the Hessian operator by computing that:

$$0 = g(X, \nabla f) \rho(Y, \nabla f) = -g(X, \nabla f) \operatorname{Hess}_f(Y, \nabla f)\,.$$

Suppose P is a point of M where $\nabla f(P) \neq 0$. Let $\{E_1, E_2, \dots, E_{m-1}, E_m = V\}$ be a local orthonormal frame which diagonalizes $\mathcal{H}_f$ and Ric where $V = \frac{\nabla f}{\|\nabla f\|}$. We may apply Equation (11.4.a) to see

$$\begin{aligned}W(E_i, E_j, E_k, V) &= \tfrac{1}{m-1}\{\rho(E_i, V)\delta_{jk} - \rho(E_j, V)\delta_{ik}\}\\ &= -\tfrac{1}{m-2}\{\rho(E_i, V)\delta_{jk} + \rho(E_j, E_k) g(E_i, V)\}\\ &\quad + \tfrac{1}{m-2}\{\rho(E_i, E_k) g(E_j, V) + \rho(E_j, V)\delta_{ik}\}\\ &\quad + \tfrac{1}{(m-1)(m-2)} \tau\{g(E_i, V)\delta_{jk} - g(E_j, V)\delta_{ik}\}\\ &= \tfrac{1}{(m-1)(m-2)}\{-\rho(E_i, V)\delta_{jk} + \rho(E_j, V)\delta_{ik}\}\\ &\quad \tfrac{1}{m-2}\{-\rho(E_j, E_k) g(E_i, V) + \rho(E_i, E_k) g(E_j, V)\}\\ &\quad + \tfrac{1}{(m-1)(m-2)} \tau\{g(E_i, V)\delta_{jk} - g(E_j, V)\delta_{ik}\}\,.\end{aligned}$$

This shows that if $1 \leq i \leq m-1$, then

$$W(V, E_i, E_i, V) = -\tfrac{1}{(m-1)(m-2)}\rho(V,V) - \tfrac{1}{m-2}\rho(E_i, E_i) + \tfrac{1}{(m-1)(m-2)}\tau\,.$$

Since the Weyl tensor vanishes, $\rho(E_i, E_i) = \frac{1}{m-1}(\tau - \rho(V,V))$. We may now show that the level sets $f^{-1}(c)$ are totally umbilical hypersurfaces of (M, g) by computing:

$$\mathrm{Hess}_f(E_i, E_i) = \lambda + \tfrac{1}{m-1}(\rho(V,V) - \tau) = \tfrac{1}{m-1}(\Delta f - \mathrm{Hess}_f(V,V))\,.$$

Since $V = \frac{\nabla f}{\|\nabla f\|}$ is a geodesic vector field, M can be written locally as a twisted product of the form $(a,b) \times N$ with metric $dt^2 + \phi^2 g_N$ where ϕ is a smooth function on $(a,b) \times N$ (see Ponge and Reckziegel [121]). Moreover, since ∇f is an eigenvector of Ric, the Ricci tensor is block diagonal with respect to the twisted product decomposition. Thus, the twisted product structure reduces to a warped product $(M, g) = ((a,b) \times N, dt^2 + \psi(t)^2 g_N)$ (see Fernández-López et al. [66]). Because $\mathcal{M}$ is locally conformally flat, it follows that (N, g_N) has constant sectional curvature (see Brozos-Vázquez, García-Río and Vázquez-Lorenzo [24]). □

The complete locally conformally flat gradient shrinking Ricci solitons has been classified. If $\mathcal{M}$ is compact, then Derdzinski [54] and Eminenti, La Nave and Mantegazza [60] showed that the only possibilities are the standard sphere or one of its quotients. If $\mathcal{M}$ is complete with non-negative Ricci curvature and if $\|R\|$ has at most exponential growth, then Ni and Wallach [107] showed the soliton must be S^m, $\mathbb{R}^m$, $\mathbb{R} \times S^{m-1}$ or one of their quotients. Cao, Wang and Zhang [39] imposed the weaker assumption that the Ricci curvature is bounded from below to derive the same conclusion. Petersen and Wylie [118] showed this conclusion followed from the integral estimate $\int_M \|\rho\|^2 e^{-f} < \infty$ where f is any potential function of the gradient shrinking Ricci soliton. Finally, Munteanu and Sesum [106] completed the classification by showing that this integral estimate holds if the Weyl tensor vanishes.

Let $\mathcal{M}$ be a complete gradient steady Ricci soliton. Bryant [27] proved that there exists, up to scaling, a unique complete rotationally symmetric gradient Ricci soliton on $\mathbb{R}^m$ in addition to the trivial Gaussian steady soliton. Cao and Chen [38] proved that these are the only possibilities under the assumption that $\mathcal{M}$ is locally conformally flat. We use work of Kotschwar [88] and Bryant [27] to establish the following result.

Theorem 11.31 *Let $\mathcal{M} = (M, g, f)$ be an m-dimensional simply connected complete locally conformally flat gradient shrinking or steady Ricci soliton. Then, after rescaling, one has the following.*

1. *If $\mathcal{M}$ is shrinking then $\mathcal{M}$ is isometric to S^m, $\mathbb{R}^m$ or $\mathbb{R} \times S^{m-1}$.*
2. *If $\mathcal{M}$ is steady then $\mathcal{M}$ is isometric to $\mathbb{R}^m$ or the Bryant soliton.*

Proof. Let P be a point of M such that $(\nabla f)(P) \neq 0$. By Lemma 11.30, one has a local decomposition of the form $(M, g) = ((-\epsilon, \epsilon) \times N, dt^2 + \psi(t)^2 g_N)$ where (N, g_N) is a Riemannian manifold of constant sectional curvature. We showed in Section 11.1 that shrinking and steady solitons are ancient solutions of the Ricci flow. Chen [42] showed that any complete ancient solution to the Ricci flow has non-negative curvature operator in dimension $m = 3$. In higher dimensions $m \geq 4$, Zhang [136] showed that a complete gradient shrinking Ricci soliton with vanishing Weyl tensor has non-negative curvature operator. Since his arguments can also be applied to the steady case, if $\mathcal{M}$ is a complete gradient shrinking or steady Ricci soliton with vanishing Weyl tensor, then the curvature operator of $\mathcal{M}$ is non-negative. Since $N = f^{-1}(c)$ is a totally umbilical hypersurface, the Gauss formula shows that (N, g_N) has non-negative sectional curvature. If the sectional curvature of (N, g_N) is zero, then N is a flat totally geodesic hypersurface and $\mathcal{M}$ is locally flat. Suppose on the other hand that (N, g_N) has constant positive curvature sectional curvature. One can use Fernández-López and García-Río [64] to see that $\mathcal{M}$ is rotationally symmetric. The desired result now follows from Bryant [27] or from Kotschwar [88] in the steady case. □

11.5 LORENTZIAN LOCALLY CONFORMALLY FLAT GRADIENT RICCI SOLITONS

We now generalize Definition 9.45 to the setting at hand.

Definition 11.32 We say that $\mathcal{M} = (M, g, f)$ is a *plane wave soliton* if we can choose local coordinates $(u, v, x^1, \dots, x^m)$ so that $f(u, x^1, \dots, x^m) f_0(u)$ where $f_0''(u) = -\rho_{uu} = m\, a(u)$ and so that $g = 2du \circ dv + H(u, x^1, \dots, x^m)du^2 + (dx^1)^2 + \cdots + (dx^m)^2$ where

$$H(u, x^1, \dots, x^m) = a(u)((x^1)^2 + \cdots + (x^m)^2) + b_1(u)x^1 + \cdots + b_m x^m + c(u)\,.$$

Throughout this section, let (M, g) be a Lorentzian manifold of dimension $m + 2$. We first examine the local structure of locally conformally flat Lorentzian gradient Ricci solitons.

Theorem 11.33 *Let (M, g, f) be a locally conformally flat Lorentzian gradient Ricci soliton.*

1. *In a neighborhood of any point where $\|\nabla f\| \neq 0$, M is locally isometric to a warped product $I \times_\psi N$ with metric $\pm dt^2 + \psi^2 g_N$ where I is a real interval and (N, g_N) is a space of constant sectional curvature c.*
2. *If $\|\nabla f\| = 0$ on a non-empty open set, then (M, g) is locally isometric to a plane wave.*

We refer Brozos-Vázquez, García-Río and Gavino-Fernández [17] for the proof of Assertion 1 in Theorem 11.33; it follows exactly the same lines used to prove Lemma 11.30. We will therefore focus on the proof of Assertion 2 and assume that $\|\nabla f\| = 0$. If $\mathcal{M}$ is Riemannian, then the holonomy group acts completely reducibly so the tangent bundle may be decomposed as the direct sum of subbundles upon which the holonomy group acts irreducibly. However, if $\mathcal{M}$ has indefinite signature, then situation is more delicate. Indecomposable but not irreducible Lorentzian manifolds admit a parallel degenerate line field $\mathfrak{D}$, and the curvature of such a manifold satisfies the following identities (see, for example, Derdzinski and Roter [55])

$$R(\mathfrak{D},\mathfrak{D}^{\perp},\cdot,\cdot)=0,\quad R(\mathfrak{D},\mathfrak{D},\cdot,\cdot)=0,\quad R(\mathfrak{D}^{\perp},\mathfrak{D}^{\perp},\mathfrak{D},\cdot)=0\,.$$

Definition 11.34 A Lorentzian manifold $\mathcal{M}$ is said to be a *pr-wave* if $\mathcal{M}$ admits a parallel degenerate line field $\mathcal{D}$ and if $R(\mathfrak{D}^{\perp},\mathfrak{D}^{\perp},\cdot,\cdot)=0$. One says that $\mathcal{M}$ has an *isotropic Ricci tensor* if the image of the Ricci tensor is totally isotropic. Leistner [101] showed that a pr-wave with isotropic Ricci tensor is a pp-wave. The local form of a pp-wave can be described as follows. There exist local coordinates $(u,v,x^1,\dots,x^m)$ and a smooth function $H(u,x^1,\dots,x^m)$ so that the Lorentzian metric is given by

$$g=2du\circ dv+H(u,x^1,\dots,x^m)du^2+(dx^1)^2+\dots+(dx^m)^2\,.$$

Lemma 11.35 Let $\mathcal{M}=(M,g,f)$ be an isotropic locally conformally flat Lorentzian gradient Ricci soliton. Then $\mathcal{M}$ is steady and locally a pp-wave.

Proof. By assumption, $\|\nabla f\|=0$. We will establish the Lemma by showing that ∇f spans a parallel null line field $\mathcal{D}$ and that $R(\mathfrak{D}^{\perp},\mathfrak{D}^{\perp},\cdot,\cdot)=0$. Set $V=\nabla f$. Since V is a null vector, we may decompose V in the form $V=S+T$ where $g(S,S)=\frac{1}{2}, g(T,T)=-\frac{1}{2}$, and $g(S,T)=0$. Let $U:=S-T$. Then $g(U,U)=0$ and $g(U,V)=g(S,S)-g(T,T)=1$. Extend $\{U,V\}$ to a local pseudo-orthonormal frame $\{U,V,E_1,\dots,E_m\}$. Since the Weyl tensor is vanishing, one has for any vector field Z that:

$$\begin{aligned}R(Z,E_i,E_j,V)&=\tfrac{1}{m+1}\{-\rho(Z,V)\delta_{ij}+\rho(E_i,V)g(Z,E_j)\}\\&=\tfrac{1}{m(m+1)}\tau\{g(Z,V)\delta_{ij}-g(E_i,V)g(Z,E_j)\}\\&\quad-\tfrac{1}{m}\{\rho(Z,V)\delta_{ij}+\rho(E_i,E_j)g(Z,V)\}\\&\quad+\tfrac{1}{m}\{\rho(Z,E_j)g(E_i,V)+\rho(E_i,V)g(Z,E_j)\}\,.\end{aligned}\tag{11.5.a}$$

By Lemma 11.5, $\nabla\tau=2\,\mathrm{Ric}(\nabla f)$ and $\tau+\|\nabla f\|^2-2\lambda f$ is constant. Since we have assumed that $\|\nabla f\|^2=0$, we may conclude that $\mathrm{Ric}(\nabla f)=\lambda\nabla f$. Consequently,

$$\rho(V,V)=0,\quad\rho(U,V)=\lambda,\quad\rho(V,E_i)=0\quad\text{for}\quad 1\le i\le m\,.$$

We apply Equation (11.5.a) to compute $R(U,E_i,E_j,V)$ and see that:

$$R(U,E_i,E_j,V)=-\tfrac{1}{m+1}\lambda\delta_{ij}=\tfrac{1}{m(m+1)}\tau\delta_{ij}-\tfrac{1}{m}\lambda\delta_{ij}-\tfrac{1}{m}\rho(E_i,E_j)\,.$$

This shows that $\rho(E_i,E_j)=\frac{1}{m+1}(\tau-\lambda)\delta_{ij}$ so $\tau=(m+2)\lambda$ Since the scalar curvature is constant, $0=\nabla\tau=2\,\mathrm{Ric}(V)=2\lambda V$. We now conclude that $\lambda=0$ and $\tau=0$. Therefore, the only possibly non-zero Ricci component is $\rho(U,U)$. Consequently, $\mathcal{M}$ is a steady gradient Ricci soliton with nilpotent Ricci operator.

Since $\lambda=0$, we may use gradient Ricci soliton Equation (11.1.b) to conclude that $\mathcal{H}_f=-\mathrm{Ric}$. Since $\mathrm{Ric}(V)=0$, $\nabla_V V=0$. This shows that V is a geodesic vector field. Since

(M, g, f) is a steady gradient Ricci soliton, the computations of Ricci operator performed above show that $\mathcal{H}_f(U) = -\operatorname{Ric}(U) = -\rho(U, U)V$, that $\mathcal{H}_f(V) = -\operatorname{Ric}(V) = 0$, and that $\mathcal{H}_f(E_i) = -\operatorname{Ric}(E_i) = 0$. Let σ be the 1-form so that $\sigma(U) := -\rho(U, U)$, $\sigma(V) = 0$ and $\sigma(E_i) = 0$ for all $1 \le i \le m$. We then have that

$$\nabla_X \nabla f = \operatorname{Hes}_f(X) = \sigma(X)\nabla f\,.$$

This shows that $\mathfrak{D} := \operatorname{span}\{\nabla f\}$ is a parallel degenerate line on $\mathcal{M}$. Furthermore, since the Weyl tensor vanishes, we may use our computation of the Ricci tensor to see that $\mathcal{M}$ is a pr-wave by verifying:

$$R(\mathfrak{D}^\perp, \mathfrak{D}^\perp, \cdot, \cdot) = 0\,.$$

Moreover, since the Ricci tensor is isotropic, this implies, as noted above, that $\mathcal{M}$ is a pp-wave. Note that although (M, g) is a pp-wave, in general ∇f is not parallel. □

Before completing the proof of Assertion 2 of Theorem 11.33, we analyze the existence of gradient Ricci solitons on pp-waves. We recall the notation of Definition 11.34. Let indices i, j range from 1 through m. To simplify the notation, let $\partial_u = \frac{\partial}{\partial_u}$, $\partial_v = \frac{\partial}{\partial_v}$, and $\partial_i = \frac{\partial}{\partial_{x^i}}$. One has:

$$\nabla_{\partial_u}\partial_u = \tfrac{1}{2}\partial_u H\,\partial_v - \tfrac{1}{2}\sum_i \partial_i H\,\partial_i \quad \text{and} \quad \nabla_{\partial_u}\partial_i = \tfrac{1}{2}\partial_i H\,\partial_v\,.$$

Consequently, ∂_v is a parallel null vector. The possibly non-vanishing components of the curvature tensor and the Ricci tensor are given (up to the usual symmetries) by

$$R_{uiuj} = -\tfrac{1}{2}\partial^2_{ij} H \quad \text{and} \quad \rho_{uu} = -\tfrac{1}{2}\sum_i \partial^2_{ii} H\,. \tag{11.5.b}$$

This shows that τ vanishes. Therefore, a pp-wave $\mathcal{M}$ is Einstein (and, therefore, Ricci flat) if and only if the space-Laplacian of the defining function H vanishes identically. The existence of gradient Ricci solitons is given by the following result of Brozos-Vázquez, García-Río and Gavino-Fernández [17].

Theorem 11.36 *Let (M, g) be a pp-wave. Then $\mathcal{M} = (M, g, f)$ is a non-trivial gradient Ricci soliton if and only if $\mathcal{M}$ is steady and if there exist constants κ_i so that*

$$f(u, x^1, \dots, x^m) = f_0(u) + \sum_{i=1}^m \kappa_i x^i \quad \text{and} \quad f_0''(u) = -\rho_{uu} - \tfrac{1}{2}\sum_{i=1}^m \kappa_i \partial_i H(u, x^1, \dots, x^m)\,.$$

One says (see, for example, Candela, Flores and Sánchez [36]) that a pp-wave $\mathcal{M}$ is a *plane wave* if

$$H(u, x^1, \dots, x^m) = \sum_{i,j} a_{ij}(u) x^i x^j\,.$$

Note that any plane wave is a isotropic steady gradient Ricci soliton for a potential function f given by $f(u, v, x^1, \dots, x^m) = f_0(u)$ where $f_0''(u) = -\rho_{uu} = \frac{1}{2}\left(\sum_i a_{ii}(u)\right)$. Moreover, Candela, Flores and Sánchez [36] showed that plane waves are geodesically complete and therefore since ∇f is a geodesic vector field, it follows that ∇f is complete. We also refer to Theorem 9.46.

Proof of Assertion 2 of Theorem 11.33. Let $\mathcal{M} = (M, g, f)$ be an isotropic locally conformally flat Lorentzian gradient soliton. By Lemma 11.35, (M, g) is a plane wave. Equation (11.5.b) shows that there exist smooth functions $a(u)$, $b_i(u)$, and $c(u)$ so that the function $H(u, x^1, \dots, x^m)$ of Definition 11.34 is given by:

$$H(u, x^1, \dots, x^m) = a(u) \sum_{i=1}^{m} (x^i)^2 + \sum_{i=1}^{m} b_i(u) x^i + c(u) .$$

The condition of Theorem 11.36 on the potential function becomes

$$f_0'' = -\rho_{uu} - \tfrac{1}{2} \sum_{i=1}^{m} \kappa_i b_i(u) - a(u) \sum_{i=1}^{m} \kappa_i x^i \quad \text{for} \quad \rho_{uu} = -m a(u) .$$

We differentiate this condition with respect to x^i to see that $a(u)\kappa_i = 0$ for all $1 \le i \le m$. Consequently, unless (M, g) is flat, it follows that $\kappa_i = 0$ for all i, and the potential function is given by $f(u, v, x^1, \dots, x^m) = f_0(u)$ where $f_0''(u) = -\rho_{uu} = m a(u)$. This completes the proof of Theorem 11.33. □

We note that Brozos-Vázquez, García-Río and Valle-Regueiro [23] studied locally conformally flat gradient Ricci almost solitons and generalized Theorem 11.33 by showing the non-existence of isotropic proper examples.

11.6 NEUTRAL SIGNATURE SELF-DUAL GRADIENT RICCI ALMOST SOLITONS

Throughout this section, we will be concerned with the interplay between affine surfaces and 4-dimensional geometry and report on work of Brozos-Vázquez and García-Río [16]. We begin by establishing some notational conventions.

Definition 11.37 Let $\mathcal{N} := (N, g)$ be a 4-dimensional oriented pseudo-Riemannian manifold of neutral signature. In this setting, the Weyl tensor W of Definition 11.1 is not irreducible but splits into two parts. We regard W an endomorphism of the bundle of 2-forms, $\Lambda^2(M)$. The *Hodge $\star$ operator* (see the discussion in Section 5.2 of Book II) is a map from $\Lambda^2(N)$ to $\Lambda^2(N)$ which is characterized by the property that $\alpha \wedge \star\beta = g(\alpha, \beta)\Omega$ where Ω is the oriented volume form. Since $\star^2 = \mathrm{Id}$, we may decompose $\Lambda^2(M) = \Lambda^2_+(M) \oplus \Lambda^2_-(M)$ into the ± 1 eigenvalues of $\star$; these are called the *self-dual* and the *anti-self-dual* 2-forms, respectively. This further decomposes $W = W^+ \oplus W^-$. We say that $\mathcal{N}$ is *self-dual* if $W^- = 0$ and *anti-self-dual* if $W^+ = 0$.

At this stage, we can always interchange the roles of self-dual and anti-self-dual by reversing the orientation. Since $\mathcal{M}$ is conformally flat if and only if $W = 0$, it is natural to say that $\mathcal{N}$ is *half-conformally flat* if W^+ or W^- is zero.

In Section 10.4, we discussed the geometry of Lorentzian Walker manifolds. Although there are Walker manifolds in all dimensions $m \geq 2$, we will be particularly be interested in the 4-dimensional setting. We refer to Walker [132] for more information on this subject.

Definition 11.38 A 4-dimensional *Walker manifold* is a neutral signature pseudo-Riemannian manifold $\mathcal{N} = (N, g)$ which has a 2-dimensional *null parallel distribution* $\mathfrak{D}$, i.e., the restriction of the metric tensor to $\mathfrak{D}$ is totally degenerate and $\nabla\mathfrak{D} \subset \mathfrak{D}$.

Let $\mathcal{N}$ be a 4-dimensional Walker manifold. Choose a local frame $\{u_1, u_2\}$ for $\mathfrak{D}$. Extend this to a local frame $\{u_1, u_2, v_1, v_2\}$ for TN so that $g(u_i, v_j) = \delta_{ij}$. If $\{u_1, u_2, \tilde{v}_1, \tilde{v}_2\}$ is another such frame, then $\tilde{v}_i = v_i + a_i^j u_j$. Consequently, $u_1 \wedge u_2 \wedge v_1 \wedge v_2 = u_1 \wedge u_2 \wedge \tilde{v}_1 \wedge \tilde{v}_2$ is invariantly defined. If we change the basis for $\mathfrak{D}$ setting $\tilde{u}_i = b_i^j u_j$ and if $\tilde{b}_i^j$ is the inverse matrix, then we may take $\tilde{v}_i = \tilde{b}_i^j v_j$ to see that the orientation is unchanged. We refer to Derdzinski [53] for further details. We fix this orientation for the remainder of this section. There exist *Walker coordinates* (x^1, x^2, y_1, y_2) where the metric tensor has the form

$$g = 2dx^i \circ dy_i + a_{ij}\, dx^i \circ dx^j \quad \text{for} \quad a_{ij} = a_{ij}(x^1, x^2, y_1, y_2)\,.$$

In such coordinates (see, for example, Brozos-Vázquez et al. [21]), the two-form $dy_1 \wedge dy_2$ in the null parallel distribution is self-dual.

Walker 4-dimensional manifolds arise naturally from the geometry of affine surfaces using the modified Riemannian extension. Although we will be primarily interested in the case of affine surfaces, it is convenient to present the definition in complete generality since, as noted in Remark 9.19, the metrics of Example 9.18 arise in this fashion.

Definition 11.39 Let $\mathcal{M} = (M, \nabla)$ be an affine manifold. If $(x^1, \dots, x^m)$ are local coordinates on M, let $(y_1, \dots, y_m)$ be the dual coordinates on the cotangent bundle so that if ω is a 1-form, then $\omega = y_i dx^i$. If X is a vector field on M, let $\Phi_X(\omega) = \omega(X)\omega \otimes \omega$. Let π be the natural projection from T^*M to M. Let ϕ is a symmetric 2-tensor field on M and let T and S be endomorphisms of TM. The *modified Riemannian extension* $\mathcal{N} = (T^*M, g_{\nabla,\phi,T,S,X})$ is defined by setting:

$$\begin{aligned} g_{\nabla,\phi,T,S,X} &:= dx^i \otimes dy_i + dy_i \otimes dx^i + \Phi_X \\ &\quad + \{\phi_{ij} - 2y_k \Gamma_{ij}{}^k + \tfrac{1}{2} y_r y_s (T_i^r S_j^s + T_j^r S_i^s)\} dx^i \otimes dx^j\,. \end{aligned} \tag{11.6.a}$$

We will omit ϕ from the notation if $\phi = 0$, we will omit $\{T, S\}$ from the notation if $T \otimes S = 0$, and we will omit X from the notation if $X = 0$. If $\mathcal{M}$ is a surface, then $\mathcal{N}$ is a 4-dimensional

Walker metric oriented by $dx^1 \wedge dx^2 \wedge dy_1 \wedge dy_2$. Since we have fixed the orientation, we cannot interchange roles that self-dual and anti-self-dual play.

Lemma 11.40 Let $\mathcal{M} = (M, \nabla)$ be an affine manifold.

1. $g_{\nabla,\phi,T,S,X}$ is an invariantly defined neutral signature metric on T^*M.
2. $\rho_{\nabla,jk} = \partial_{x^i}\Gamma_{jk}{}^i - \partial_{x^j}\Gamma_{ik}{}^i + \Gamma_{in}{}^i\Gamma_{jk}{}^n - \Gamma_{jn}{}^i\Gamma_{ik}{}^n$.
3. If $\mathcal{N} = (T^*M, g_{\nabla,\phi})$, then $\rho_{g_{\nabla,\phi}} = 2\{\pi^*\rho_{s,\nabla}\}$ is independent of ϕ, where $\rho_{s,\nabla}$ denotes the symmetric Ricci tensor.

Proof. There are many possible proofs of Assertion 1. We have chosen to present a combinatorial proof as we think it illustrative. For a different approach, we refer, for example, to Calviño-Louzao et al. [34]. Let $\xi = a_i\partial_{x^i} + b^j\partial_{y_j}$ be a tangent vector in T^*M. Suppose $g(\xi, \eta) = 0$ for all η. Since $g(\xi, \partial_{y_i}) = a_i$, we have $a_i = 0$ for all i. We then have $g(\xi, \partial_{x^i}) = b^i$ and, therefore, $b^i = 0$ for all i. This shows $\xi = 0$. Consequently, g is non-degenerate. Let $\mathcal{D} := \text{span}\{\partial_{y_i}\}$. Since $g(\eta, \xi) = 0$ for $\eta, \xi \in \mathcal{D}$, $\mathcal{D}$ is a null distribution of dimension m. It is then a straightforward exercise to see g has neutral signature. We complete the proof by showing g is invariantly defined. It is immediate that Φ_X and $\phi_{ij}dx^i \otimes dx^j$ are invariantly defined. Let $(\tilde{x}^1, \dots, \tilde{x}^m)$ be another choice of coordinates on M. Expand $\partial_{x^i} = a_i^j\partial_{\tilde{x}^j}$ and $dx^i = b_j^i d\tilde{x}^j$ where b is the inverse of the matrix a. Since $y_i dx^i = \tilde{y}_j d\tilde{x}^j$, $y_i = a_i^j\tilde{y}_j$.

Step 1: We express T and S relative to the new coordinate system to express $T_i^r = b_k^r\tilde{T}_\ell^k a_i^\ell$ and $S_j^s = b_u^s\tilde{S}_n^u a_j^n$. We use the fact that a and b are inverse matrices to see

$$
\begin{aligned}
y_r y_s T_i^r S_j^s dx^i \otimes dx^j &= a_r^\alpha a_s^\beta b_k^r a_i^\ell b_u^s a_j^n b_\gamma^i b_\mu^j \tilde{y}_\alpha\tilde{y}_\beta\tilde{T}_\ell^k\tilde{S}_n^u d\tilde{x}^\gamma \otimes d\tilde{x}^\mu \\
&= a_i^\ell b_\gamma^i\ b_\mu^j a_j^n\ a_r^\alpha b_k^r\ a_s^\beta b_u^s\ \tilde{y}_\alpha\tilde{y}_\beta\tilde{T}_\ell^k\tilde{S}_n^u d\tilde{x}^\gamma \otimes d\tilde{x}^\mu \\
&= \delta_{\gamma\ell}\delta_{\mu n}\delta_{k\alpha}\delta_{u\beta}\tilde{y}_\alpha\tilde{y}_\beta\tilde{T}_\ell^k\tilde{S}_n^u d\tilde{x}^\gamma \otimes d\tilde{x}^\mu = \tilde{y}_\alpha\tilde{y}_\beta\tilde{T}_\gamma^\alpha\tilde{S}_\mu^\beta d\tilde{x}^\gamma \otimes d\tilde{x}^\mu .
\end{aligned}
$$

Step 2: We complete the proof by examining $dx^i \otimes dy_i + dy_i \otimes dx^i - 2y_k\Gamma_{ij}{}^k$. We compute:

$$
\begin{aligned}
\nabla_{\partial_{x^i}}\partial_{x^j} &= a_i^\alpha\nabla_{\partial_{\tilde{x}^\alpha}}a_j^\beta\partial_{\tilde{x}^\beta} = a_i^\alpha\{a_j^\beta\tilde{\Gamma}_{\alpha\beta}{}^\gamma + \partial_{\tilde{x}^\alpha}a_j^\gamma\}\partial_{\tilde{x}^\gamma} \\
&= a_i^\alpha b_\gamma^k\{a_j^\beta\tilde{\Gamma}_{\alpha\beta}{}^\gamma + \partial_{\tilde{x}^\alpha}a_j^\gamma\}\partial_{x^k},
\end{aligned}
$$

$$
\Gamma_{ij}{}^k = a_i^\alpha b_\gamma^k\{a_j^\beta\tilde{\Gamma}_{\alpha\beta}{}^\gamma + \partial_{\tilde{x}^\alpha}a_j^\gamma\},
$$

$$
dy_i = d(a_i^\beta\tilde{y}_\beta) = a_i^\beta d\tilde{y}_\beta + \partial_{\tilde{z}^\gamma}(a_i^\beta)\tilde{y}_\beta d\tilde{x}^\gamma,
$$

$$
\begin{aligned}
&dx^i \otimes \{dy_i - y_k\Gamma_{ij}{}^k dx^j\} \\
&\quad = b_v^i d\tilde{x}^v \otimes \{a_i^\beta d\tilde{y}_\beta + \partial_{\tilde{z}^\gamma}(a_i^\beta)\tilde{y}_\beta d\tilde{x}^\gamma - a_k^u\tilde{y}_u(a_i^\alpha b_\gamma^k\{a_j^\beta\tilde{\Gamma}_{\alpha\beta}{}^\gamma + \partial_{\tilde{x}^\alpha}a_j^\gamma)b_w^j d\tilde{x}^w\} \\
&\quad = b_v^i a_i^\beta d\tilde{x}^v \otimes d\tilde{y}_\beta - b_v^i a_k^u a_i^\alpha b_\gamma^k a_j^\beta b_w^j\tilde{\Gamma}_{\alpha\beta}{}^\gamma\tilde{y}_u d\tilde{x}^v \otimes d\tilde{x}^w \\
&\qquad + b_v^i\partial_{\tilde{z}^\gamma}(a_i^\beta)\tilde{y}_\beta d\tilde{x}^v \otimes d\tilde{x}^\gamma - b_v^i a_k^u a_i^\alpha b_\gamma^k b_w^j\partial_{\tilde{x}^\alpha}(a_j^\gamma)\tilde{y}_u d\tilde{x}^v \otimes d\tilde{x}^w
\end{aligned}
$$

$$
\begin{aligned}
&= d\tilde{x}^v \otimes d\tilde{y}_v - \tilde{\Gamma}_{vw}{}^u \tilde{y}_u d\tilde{x}^v \otimes d\tilde{x}^w \\
&\quad + b_v^i \partial_{\tilde{z}^\gamma}(a_i^\beta)\tilde{y}_\beta d\tilde{x}^v \otimes d\tilde{x}^\gamma - b_w^j \partial_{\tilde{x}^v}(a_j^u)\tilde{y}_u d\tilde{x}^v \otimes d\tilde{x}^w \\
&= d\tilde{x}^v \otimes d\tilde{y}_v - \tilde{\Gamma}_{vw}{}^u \tilde{y}_u d\tilde{x}^v \otimes d\tilde{x}^w + b_v^i \partial_{\tilde{z}^w}(a_i^u)\tilde{y}_u\{d\tilde{x}^v \otimes d\tilde{x}^w - d\tilde{x}^w \otimes d\tilde{x}^v\}.
\end{aligned}
$$

The term $b_v^i \partial_{\tilde{z}^w}(a_i^u)\tilde{y}_u\{d\tilde{x}^v \otimes d\tilde{x}^w - d\tilde{x}^w \otimes d\tilde{x}^v\}$ vanishes when we symmetrize. This shows that $dx^i \otimes \{dy_i - y_k \Gamma_{ij}{}^k\} + \{dy_i - y_k \Gamma_{ij}{}^k \otimes dx^i\}$ is invariantly defined and establishes Assertion 1. We will omit the proof of Assertion 2 and Assertion 3 as it is a straightforward calculation. □

We can use this construction to characterize self-dual 4-dimensional Walker manifolds. We refer to Calviño-Louzao et al. [34] and Díaz-Ramos, García-Río and Vázquez-Lorenzo [57] for the proof of the following result.

Theorem 11.41 *Let $\mathcal{N} = (N, g)$ be a 4-dimensional Walker manifold. Then $\mathcal{N}$ is self-dual if and only if $\mathcal{N}$ is locally isometric to $(T^*M, g_{\nabla,\phi,T,S,X})$ for some $(M, \nabla, \phi, T, S, X)$.*

We adopt the notation of Definition 9.2 to define the Ricci tensor ρ and the symmetric Ricci tensor ρ_s. If h is a smooth function on M, we say that (M, ∇, h) is an *affine gradient Ricci soliton* if f satisfies the equation $\operatorname{Hess}_h + 2\rho_s = 0$. We will be interested in inducing Ricci solitons on the cotangent bundle from suitable structures on the base. In Section 11.6.5, we will examine the isotropic case and in particular the structures $(T^*M, g_{\nabla,\phi}, \pi^*h)$ where $g_{\nabla,\phi}$ is defined by Equation (11.6.a). We will then use these results in Section 11.6.6 to establish the following result.

Lemma 11.42 If (M, ∇) is an affine surface, then (M, ∇, h) is an affine gradient Ricci soliton if and only if $\mathcal{N} := (T^*M, g_{\nabla,\phi}, \pi^*h)$ is a gradient Ricci soliton. In this setting, $\mathcal{N}$ is self-dual, the soliton is steady, $\|\nabla \pi^* h\| = 0$, $\tau_{\mathcal{N}} = 0$, and $\mathcal{W}^+$ is nilpotent.

It is worth presenting a few examples to illustrate this phenomena. They arose out of our study of homogeneous 2-dimensional affine surfaces (Brozos-Vázquez, García-Río and Gilkey [20]). If $\mathcal{M}$ is an affine surface, let $\mathfrak{A}(\mathcal{M})$ be the set of affine gradient Ricci solitons on $\mathcal{M}$. If $\mathfrak{A}(\mathcal{M})$ is non-empty, then $\mathfrak{A}(\mathcal{M})$ is an affine space; if $f_i \in \mathfrak{A}(\mathcal{M})$, then for any $t \in \mathbb{R}$, we form the 1-parameter family $tf_1 + (1-t)f_2 \in \mathfrak{A}(\mathcal{M})$.

We recall the notation of Section 9.6.1. Let $\mathcal{M} = (M, \nabla)$ be an affine surface. Then $\mathcal{M}$ is a *Type $\mathcal{A}$ geometry* if $M = \mathbb{R}^2$ and if the Christoffel symbols are constant. Similarly, $\mathcal{M}$ is a *Type $\mathcal{B}$ geometry* if $M = \mathbb{R}^+ \times \mathbb{R}$ and if the Christoffel symbols have the form $\Gamma_{ij}{}^k = (x^1)^{-1} C_{ij}{}^k$ for $C_{ij}{}^k$ constant. These two classes are not exclusive; a Type $\mathcal{B}$ geometry is locally isomorphic to a Type $\mathcal{A}$ geometry if and only if $C_{12}{}^1 = C_{22}{}^1 = C_{22}{}^2 = 0$. The importance of these two classes of surfaces lies in the classification result of Opozda [111] given in Theorem 9.48 previously. The affine gradient Ricci solitons are known for these geometries. We refer to Brozos-Vázquez, García-Río and Gilkey [20] for the proof of the following two results.

Theorem 11.43 *Suppose that $\mathcal{M}$ is a Type $\mathcal{A}$ geometry with $\Gamma_{11}{}^2 = 0$ and $\Gamma_{12}{}^2 = 0$.*

1. $\rho = \{\Gamma_{12}{}^1(\Gamma_{22}{}^2 - \Gamma_{12}{}^1) + \Gamma_{11}{}^1\Gamma_{22}{}^1\}dx^2 \otimes dx^2$.
2. $f \in \mathfrak{A}(\mathcal{M})$ *if and only if* $f(x^1, x^2) = \xi(x^2)$ *where* $\xi'' - \Gamma_{22}{}^2\xi' + \rho_{22} = 0$.
3. *Up to linear equivalence, these are the only Type* $\mathcal{A}$ *geometries with* $\mathfrak{A}(\mathcal{M}) \neq \emptyset$.

We now turn to the Type $\mathcal{B}$ geometries. Let $(a, c) \neq (0, 0)$. Let $\tilde{c} \in \mathbb{R}$. Set

$$\begin{array}{llll}
\mathcal{P}^{\pm}_{a,c}: & C_{11}{}^1 = \frac{1}{2}\left(a^2 + 4a \mp 2c^2 + 2\right), & C_{11}{}^2 = c, & C_{12}{}^1 = 0, \\
& C_{12}{}^2 = \frac{1}{2}\left(a^2 + 2a \mp 2c^2\right), & C_{22}{}^1 = \pm 1, & C_{22}{}^2 = \pm 2c, \\
\mathcal{Q}_{\tilde{c}}: & C_{11}{}^1 = 0, & C_{11}{}^2 = \tilde{c}, & C_{12}{}^1 = 1, \\
& C_{12}{}^2 = 0, & C_{22}{}^1 = 0, & C_{22}{}^2 = 1.
\end{array}$$

Theorem 11.44 *Let* $\mathcal{M}$ *be a Type* $\mathcal{B}$ *affine surface which is not locally isomorphic to a Type* $\mathcal{A}$ *surface.*

1. $\mathfrak{A}(\mathcal{M})$ *is non-empty if and only if* $\mathcal{M}$ *is linearly isomorphic to* $\mathcal{P}^{\pm}_{a,c}$ *or to* $\mathcal{Q}_{\tilde{c}}$.
2. *If* $\mathcal{M}$ *is isomorphic to* $\mathcal{P}^{\pm}_{0,c}$ *or to* $\mathcal{Q}_{\tilde{c}}$, *then* $\rho_{\mathcal{M}}$ *is alternating and* $\mathfrak{A}(\mathcal{M}) = \mathbb{R}$.
3. *If* $\mathcal{M} = \mathcal{P}^{\pm}_{-2,0}$, *then* $\mathfrak{A}(M) = \{-2\log(x^1) + c_1x^2 + c_0\}$ *for* $c_i \in \mathbb{R}$.
4. *If* $\mathcal{M} = \mathcal{P}^{-}_{-\frac{1}{2},c}$, *then* $\mathfrak{A}(\mathcal{M}) = \{-\frac{1}{2}\log(x^1) + c_1(x^2 - \frac{3}{8}x^1) + c_0\}$ *for* $c_i \in \mathbb{R}$.
5. *If* $\mathcal{M}$ *is as in Assertion 1 but does not satisfy Assertions 2–4, then there exists* $a \neq 0$ *so that* $\mathfrak{A}(\mathcal{M}) = \{a\log(x^1) + c_0\}$ *for* $c_0 \in \mathbb{R}$.

Note that the examples in Assertion 2 of Theorem 11.44 are "trivial" in the sense that the Ricci tensor of $\mathcal{M}$ is alternating and, consequently, $(T^*M, g_{\nabla,\phi})$ is Ricci flat. The other examples of Theorem 11.44 are non-trivial as are the examples of Theorem 11.43.

Two connections ∇ and $\tilde{\nabla}$ are said to be *projectively equivalent* if there exists a smooth 1-form ω so $\nabla_X Y - \tilde{\nabla}_X Y = \omega(X)Y + \omega(Y)X$ or, equivalently, the unparameterized geodesics of ∇ and $\tilde{\nabla}$ agree. We say that ∇ and $\tilde{\nabla}$ are *strongly projectively equivalent* if $d\omega = 0$. Finally, we say that ∇ is *projectively flat* if ∇ is strongly projectively equivalent to a flat connection. Note that any Type $\mathcal{A}$ surface is projectively flat and that the surfaces $\mathcal{P}^{\pm}_{0,c}$ and $\mathcal{Q}_{\tilde{c}}$ are not projectively flat (see Brozos-Vázquez et al. [22]). Work of Calviño-Louzao et al. [34], and Díaz-Ramos, García-Río and Vázquez-Lorenzo [57] shows that (T^*M, g_∇) is locally conformally flat if and only if (M, ∇) is projectively flat. Furthermore, by varying ϕ, we can construct examples where $(T^*M, g_{\nabla,\phi})$ is no longer conformally flat even if (M, ∇) is projectively flat.

Chen and Wang [43] showed that all non-trivial self-dual Riemannian gradient Ricci solitons are locally conformally flat. We will establish the following result in Section 11.6.2.

Theorem 11.45 Let $\mathcal{N} = (N, g, f)$ be a 4-dimensional half-conformally flat gradient Ricci almost soliton of arbitrary signature with $\|\nabla f\| \neq 0$. Then $\mathcal{N}$ is locally isometric to a warped product $I \times_\varphi U$ where $\mathcal{U}$ is a 3-dimensional manifold of constant sectional curvature. Hence, $\mathcal{N}$ is locally conformally flat.

Dunajski and West [58] showed that the geometry of (anti)-self-dual conformal structures is much richer in neutral signature $(2,2)$. The isotropic case where $\|\nabla f\| = 0$ is qualitatively different as there are many examples which are not locally conformally flat. We will always assume the (almost) soliton is non-trivial, i.e., $\nabla f \neq 0$. All isotropic self-dual gradient Ricci almost solitons are steady and they are locally isomorphic to the cotangent bundle of an affine surface with a suitable generalized Riemannian extension. They form a special class of Walker metrics (see García-Río et al. [69]). This is a neutral signature phenomena not present in the Riemannian setting.

We show presently that any self-dual gradient Ricci almost soliton which is not locally conformally flat is isotropic. Moreover, there are steady traceless κ-Einstein solitons, i.e., they satisfy the equation $\mathrm{Hess}_f + \rho = \lambda\, g$ for a function $\lambda = \kappa\,\tau + \mu$ with $\mu = 0$ and $\kappa = \frac{1}{4}$. The following result will be established in Section 11.6.4.

Theorem 11.46 *Let $\mathcal{N} = (N, g, f)$ be a 4-dimensional half-conformally flat proper gradient Ricci almost soliton with $\|\nabla f\| = 0$ and $\nabla f \neq 0$. Then $\mathcal{N}$ is locally isomorphic to $(T^*M, g_{\nabla,\phi,T,\mathrm{Id}}, \pi^* h)$ where $\mathcal{M} = (M, \nabla)$ is an affine surface, $\lambda = \frac{3}{2} C e^{-f}$, $T = C e^{-h}\,\mathrm{Id}$, $\phi = \frac{2}{C} e^{h}(\mathrm{Hess}_h + 2\rho_{\nabla,s})$, and $C \in \mathbb{R}$.*

Section 11.6.5 is devoted to examining the isotropic case. The results of that section will be used in Section 11.6.6 to prove the following result. It provides the converse to Lemma 11.42 and is the main result of this section.

Theorem 11.47 *Let $\mathcal{N} = (N, g, f)$ be a 4-dimensional half-conformally flat non-trivial gradient Ricci soliton with $\|\nabla f\| = 0$ and $\nabla f \neq 0$. Then $\mathcal{N}$ is locally isomorphic to $(T^*M, g_{\nabla,\phi}, \pi^* h)$ for some affine surface $\mathcal{M} = (M, \nabla)$ where h is an affine gradient Ricci soliton on $\mathcal{M}$.*

Work of Brozos-Vázquez, García-Río and Gavino-Fernández [17] shows that any isotropic locally conformally flat Lorentzian gradient Ricci soliton is a plane wave. Locally conformally flat Lorentzian gradient Ricci almost solitons have been investigated by Brozos-Vázquez, García-Río and Valle-Regueiro [23]. The isotropic case reduces to that of steady gradient Ricci solitons whose underlying geometry corresponds to a plane wave. We also refer to work of Afifi [1].

Theorem 11.45 and Theorem 11.46 show that gradient Ricci solitons and gradient Ricci almost solitons behave quite diferently in the isotropic setting. The potential function of any self-dual gradient Ricci soliton is the pullback of a solution to the affine Ricci soliton equation and is unrelated to the deformation ϕ. On the other hand, any smooth function on the affine surface $\mathcal{M}$ pulls back to define an almost soliton where ϕ and T must be chosen appropriately.

11.6.1 ALGEBRAIC PRELIMINARIES.

Definition 11.48 Let $(V, \langle\cdot,\cdot\rangle)$ be a 4-dimensional neutral signature inner product space. A basis $\mathcal{B} = \{e_1, e_2, e_3, e_4\}$ is said to be an *orthonormal basis* if the non-zero components of $\langle\cdot,\cdot\rangle$ are given by $\langle e_i, e_j\rangle = \epsilon_i$ for $\epsilon_i = \pm 1$. Similarly, $\mathcal{B}$ is said to be a *pseudo-orthonormal basis* or a *hyperbolic basis* if the non-zero components of $\langle\cdot,\cdot\rangle$ are given by $\langle e_1, e_3\rangle = \langle e_2, e_4\rangle = 1$. A tensor $R \in \otimes^4 V^*$ is said to be an *algebraic curvature tensor* if R satisfies the $\mathbb{Z}_2$-symmetries of the Riemann curvature tensor and if R also satisfies the second Bianchi identity. We use Definition 11.1 to define the associated Weyl conformal curvature tensor and decompose $W = W_+ \oplus W_-$. We say R is a *self-dual algebraic curvature tensor* if $W_- = 0$.

We have the following algebraic characterization of this condition.

Lemma 11.49 Let R be an algebraic curvature tensor on an oriented 4-dimensional inner product space $(V, \langle\cdot,\cdot\rangle)$ of neutral signature. The following conditions are equivalent.

1. R is self-dual.
2. If $\{e_1, e_2, e_3, e_4\}$ is any oriented orthonormal basis, if $x, y \in V$ and if $\sigma = (ijk)$ is any permutation of (234), then $W(e_1, e_i, x, y) = \operatorname{sign}(\sigma)\langle e_j, e_j\rangle\langle e_k, e_k\rangle W(e_j, e_k, x, y)$.
3. If $\{t, u, v, w\}$ is any oriented pseudo-orthonormal basis and if $x, y \in V$, then we have $W(t, v, x, y) = W(u, w, x, y)$, $W(t, w, x, y) = 0$, and $W(u, v, x, y) = 0$.

Proof. Let $\{e_1, e_2, e_3, e_4\}$ be an orthonormal basis of $(V, \langle\cdot,\cdot\rangle)$ such that $e^1 \wedge e^2 \wedge e^3 \wedge e^4$ gives the oriented volume form. We compute:

$$\Lambda^2_\pm = \operatorname{span}\{e^1 \wedge e^2 \pm \epsilon_3\epsilon_4 e^3 \wedge e^4,\ e^1 \wedge e^3 \mp \epsilon_2\epsilon_4 e^2 \wedge e^4,\ e^1 \wedge e^4 \pm \epsilon_2\epsilon_3 e^2 \wedge e^3\}\,.$$

Thus, $\mathcal{W}_- = 0$ if and only if $\mathcal{W}(e^1 \wedge e^2 - \epsilon_3\epsilon_4 e^3 \wedge e^4) = 0$, $\mathcal{W}(e^1 \wedge e^3 + \epsilon_2\epsilon_4 e^2 \wedge e^4) = 0$, and $\mathcal{W}(e^1 \wedge e^4 - \epsilon_2\epsilon_3 e^2 \wedge e^3) = 0$. Consequently, $\mathcal{W}(e_1, e_i) = \sigma_{ijk}\epsilon_j\epsilon_k\mathcal{W}(e_j, e_k)$. This proves the equivalence of Assertion 1 and Assertion 2. Suppose, next, that $\{t, u, v, w\}$ is as in Assertion 3. We construct the following oriented orthonormal basis to derive the equivalence of Assertion 2 and Assertion 3:

$$e_1 := \tfrac{1}{\sqrt{2}}(t - v),\ e_2 := \tfrac{1}{\sqrt{2}}(t + v),\ e_3 := \tfrac{1}{\sqrt{2}}(w - u),\ e_4 := \tfrac{1}{\sqrt{2}}(w + u)\,. \qquad \square$$

11.6.2 THE PROOF OF THEOREM 11.45. Let $\mathfrak{C}(X, Y, Z) = -\frac{m-2}{m-3}(\operatorname{div} W)(X, Y, Z)$ be the *Cotton tensor*. Equivalently,

$$\mathfrak{C}(X, Y, Z) = (\nabla_X \rho)(Y, Z) - (\nabla_Y \rho)(X, Z) - \tfrac{1}{2(m-1)}\{X(\tau)g(Y, Z) - Y(\tau)g(X, Z)\}\,.$$

By Lemma 11.11, the Weyl tensor of any gradient Ricci almost soliton satisfies

$$\begin{aligned}
&W(X,Y,Z,\nabla f)\\
&= -\mathfrak{C}(X,Y,Z) + \tfrac{\tau}{(m-1)(m-2)}\{g(X,Z)g(Y,\nabla f) - g(X,\nabla f)g(Y,Z)\}\\
&\quad + \tfrac{1}{(m-2)}\{\rho(Y,Z)g(X,\nabla f) - \rho(X,Z)g(Y,\nabla f)\}\\
&\quad + \tfrac{1}{(m-1)(m-2)}\{\rho(X,\nabla f)g(Y,Z) - \rho(Y,\nabla f)g(X,Z)\}.
\end{aligned} \tag{11.6.b}$$

Let $\mathcal{N} = (N, g, f)$ be a 4-dimensional half-conformally flat gradient Ricci almost soliton with $\|\nabla f\| \neq 0$. Choose the orientation so $\mathcal{N}$ is self-dual. Let P be a point of N. We must show that $\mathcal{N}$ is isometric near P to a warped product $I \times_\varphi U$ where $\mathcal{U}$ is a 3-dimensional manifold of constant sectional curvature. Since the Cotton tensor is a constant multiple of the divergence of the Weyl tensor, the self-duality condition of Assertion 2 of Lemma 11.49 applied to ∇f yields:

$$\begin{aligned}
&\tau\{g(e_i,\nabla f)e_1 - g(e_1,\nabla f)e_i\} - \{\rho(e_i,\nabla f)e_1 - \rho(e_1,\nabla f)e_i\\
&\qquad +3g(e_i,\nabla f)\operatorname{Ric}(e_1) - 3g(e_1,\nabla f)\operatorname{Ric}(e_i)\}\\
&= \sigma_{ijk}\,\epsilon_j\epsilon_k\big(\tau\{g(e_k,\nabla f)e_j - g(e_j,\nabla f)e_k\} - \{\rho(e_k,\nabla f)e_j - \rho(e_j,\nabla f)e_k\\
&\qquad +3g(e_k,\nabla f)\operatorname{Ric}(e_j) - 3g(e_j,\nabla f)\operatorname{Ric}(e_k)\}\big) \quad \text{for} \quad i,j,k \in \{2,3,4\}.
\end{aligned} \tag{11.6.c}$$

Since $\|\nabla f\| \neq 0$, we may set $E_1 := \frac{\nabla f}{\|\nabla f\|}$ and extend E_1 to an orthonormal frame $\{E_1, E_2, E_3, E_4\}$. Let $i,j,k \in \{2,3,4\}$ henceforth. By Equation (11.6.c),

$$\begin{aligned}
&-\tau g(E_1,\nabla f)g(E_i,Z) + 3\rho(E_i,Z)g(E_1,\nabla f)\\
&\quad +\rho(E_1,\nabla f)g(E_i,Z) - \rho(E_i,\nabla f)g(E_1,Z)\\
&= \sigma_{ijk}\epsilon_j\epsilon_k\{\rho(E_j,\nabla f)g(E_k,Z) - \rho(E_k,\nabla f)g(E_j,Z)\}.
\end{aligned} \tag{11.6.d}$$

If we set $Z = E_1$ in Equation (11.6.d), then we obtain $\rho(E_1, E_i) = 0$ for $2 \le i \le 4$. This shows that ∇f is an eigenvector of the Ricci operator. Next, we set $Z = E_j$ in Equation (11.6.d) to obtain

$$3\rho(E_i,E_j)g(E_1,\nabla f) = -\sigma_{ijk}\epsilon_j\epsilon_k\rho(E_k,\nabla f)g(E_j,E_j) = 0.$$

Thus, $\rho(E_i, E_j) = 0$ for all $i \neq j$. Finally, we set $Z = E_i$ in Equation (11.6.d) to obtain

$$\tau g(E_1,\nabla f)g(E_i,E_i) - 3\rho(E_i,E_i)g(E_1,\nabla f) - \rho(E_1,\nabla f)g(E_i,E_i) = 0.$$

This shows that $3\epsilon_i\rho(E_i,E_i) = \tau - \epsilon_1\rho(E_1,E_1)$. Consequently, the basis $\{E_1,\dots,E_4\}$ diagonalizes the Ricci operator. Furthermore, the Ricci operator has at most two distinct eigenvalues and, if there are two eigenvalues, then the eigenvalue of multiplicity 1 corresponds to the eigenvector E_1. We use the Ricci almost soliton Equation (11.1.b) to see that the level hypersurfaces of f are totally umbillic by computing:

$$\operatorname{Hess}_f(E_i,E_i) = \lambda g(E_i,E_i) - \rho(E_i,E_i) = \{\lambda - \tfrac{1}{3}(\tau - \epsilon_1\rho(E_1,E_1))\}g(E_i,E_i).$$

Since the 1-dimensional distribution span$\{E_1\}$ is totally geodesic, $\mathcal{N}$ decomposes locally as a twisted product $I \times_\varphi U$ (see Ponge and Reckziegel [121]). Since the Ricci tensor is diagonal, the twisted product reduces to a warped product (see Fernández-López et al. [66]). Finally, since $I \times_\varphi U$ is self-dual, it is necessarily locally conformally flat and the fiber U has constant sectional curvature (see Brozos-Vázquez, García-Río and Vázquez-Lorenzo [25]). This completes the proof of Theorem 11.45. □

11.6.3 ISOTROPIC RICCI ALMOST SOLITONS AND WALKER GEOMETRY.

Henceforth, let $\mathcal{N} = (N, g, f)$ be a 4-dimensional half-conformally flat gradient Ricci almost soliton with $\|\nabla f\| = 0$ and $\nabla f \neq 0$. In contrast with the non-isotropic case, the level hypersurfaces of the potential function are now degenerate hypersurfaces. Our first task is to show $\mathcal{N}$ admits a parallel degenerate 2-dimensional distribution. We begin by establishing the following result.

Lemma 11.50 Let $\mathcal{N} = (N, g, f)$ be a 4-dimensional half-conformally flat proper gradient Ricci almost soliton with $\|\nabla f\| = 0$ and $\nabla f \neq 0$. Then $\mathcal{N}$ is locally a Walker manifold and $\lambda = \frac{1}{4}\tau$.

Proof. We determine the structure of the Ricci operator as follows. Since $\nabla f \neq 0$ but $g(\nabla f, \nabla f) = 0$, there is a local pseudo-orthonormal frame $\mathcal{B} = \{\nabla f, U, V, T\}$. Since $g(\nabla f, \nabla f) = 0$, $0 = \nabla_X g(\nabla f, \nabla f) = 2g(\nabla_X \nabla f, \nabla f) = 2g(\nabla_{\nabla f} \nabla f, X)$ and we have that $\mathcal{H}_f(\nabla f) = 0$. The almost soliton Equation (11.1.b) now implies that $\operatorname{Ric}(\nabla f) = \lambda \nabla f$ so ∇f is an eigenvector of the Ricci operator. We apply Lemma 11.49 to the pseudo-orthonormal frame $\mathcal{B}$ to see that for any vector fields X and Y we have that

$$W(\nabla f, V, X, Y) = W(U, T, X, Y), \quad W(\nabla f, T, X, Y) = 0, \quad W(U, V, X, Y) = 0\,. \tag{11.6.e}$$

We set $Y = \nabla f$ in Equation (11.6.e). Since $\operatorname{Ric}(\nabla f) = \lambda \nabla f$, Lemma 11.49 implies that

$$0 = W(\nabla f, V, X, \nabla f) - W(U, T, X, \nabla f) = \tfrac{1}{6}(\tau - 4\lambda) g(\nabla f, X) \quad \text{for any} \quad X\,.$$

Hence, $\tau = 4\lambda$. Setting $Y = \nabla f$ in the identity $W(U, V, X, Y) = 0$ of Equation (11.6.e) and applying Lemma 11.49, we conclude that

$$0 = W(U, V, X, \nabla f) = \tfrac{1}{6}(\tau - \lambda) g(U, X) - \tfrac{1}{2}\rho(U, X) = \tfrac{1}{2}(\lambda g(U, X) - \rho(U, X)) \text{ for all } X\,.$$

This shows that $\operatorname{Ric}(U) = \lambda U$ so U is an eigenvector of the Ricci operator. Finally, if we set $X = V$ in the identity $W(\nabla f, T, X, Y) = 0$ of Equation (11.6.e) and use Lemma 11.49 once again, we have that

$$0 = W(V, Y, T, \nabla f) = -\tfrac{1}{2}\{\lambda g(Y, T) - \rho(Y, T) + \rho(V, T) g(Y, \nabla f)\} \quad \text{for all} \quad Y\,.$$

Consequently, $\rho(T,T)=\rho(\nabla f,T)=0$. We may therefore conclude that there exist smooth functions α and β on N so that relative to the basis $\mathcal{B}$, the Ricci operator has the form:

$$\mathrm{Ric}=\begin{pmatrix}\lambda & 0 & \alpha & \beta\\ 0 & \lambda & \beta & 0\\ 0 & 0 & \lambda & 0\\ 0 & 0 & 0 & \lambda\end{pmatrix}. \tag{11.6.f}$$

Let $\mathfrak{D}:=\operatorname{span}\{\nabla f, U\}$ be a 2-dimensional null distribution. We have $g(\nabla_X\nabla f,\nabla f)=0$ and $g(U,U)=0$. A similar argument shows $g(\nabla_X U,U)=0$ for all X. Because $\mathrm{Ric}(U)=\lambda U$, Equation (11.1.b) implies that $\mathcal{H}_f(U)=0$. Since $g(U,\nabla f)=0$,

$$g(\nabla_X U,\nabla f)=-g(U,\nabla_X\nabla f)=-\operatorname{Hess}_f(U,X)=0 \quad \text{for all} \quad X\,.$$

Hence, since $\mathfrak{D}^\perp=\mathfrak{D}$ and $g(\nabla_X\nabla f,\nabla f)=0$, $g(\nabla_X U,U)=0$, $g(\nabla_X U,\nabla f)=0$, and $g(U,\nabla_X\nabla f)=0$, we may conclude that $\nabla\mathfrak{D}\subset\mathfrak{D}$. Consequently, (M,g) is locally a Walker manifold. □

The choice of orientation did not play any role in our previous discussion and for that reason we used the terminology "half-conformally flat". However, a Walker manifold inherits a natural orientation and thus the self-dual and the anti-self-dual conditions are not interchangeable in this context. Díaz-Ramos, García-Río and Vázquez-Lorenzo [57] noted that if the self-dual Weyl curvature W^+ of a Walker manifold vanishes, then $\tau=0$ and, consequently, by Lemma 11.50, $\lambda=0$. This proves the following result.

Lemma 11.51 Let $\mathcal{N}=(N,g,f)$ be a 4-dimensional anti-self-dual Walker manifold which is a non-trivial isotropic gradient Ricci almost soliton. Then $\mathcal{N}$ is a steady gradient Ricci soliton.

For the remainder of this section, we consider the cases of proper gradient Ricci almost solitons and gradient Ricci solitons separately.

11.6.4 THE PROOF OF THEOREM 11.46.

Theorem 11.46 will follow from Lemma 11.51 and from the following result.

Lemma 11.52 Let $\mathcal{N}=(N,g,f)$ be a 4-dimensional Walker manifold which is a self-dual isotropic gradient Ricci almost soliton. Then (N,g) is locally isomorphic to $(T^*M, g_{\nabla,\phi,T,\mathrm{Id}})$ of some affine surface (M,∇). Assume that $\lambda\neq 0$.

1. The potential function satisfies $f=\pi^*h$ for some smooth function h on M and f is related with the soliton function by $\lambda=\frac{3}{2}Ce^{-f}$ for a constant C.
2. We have $T=Ce^{-h}\,\mathrm{Id}$ and $\phi=\frac{2}{C}e^{h}(\operatorname{Hess}_h+2\rho_{\nabla,s})$.

Proof. Work of Calviño-Louzao et al. [34] shows that a 4-dimensional Walker manifold is self-dual if and only if it is locally isometric to the cotangent bundle T^*M of an affine surface (M, ∇), with metric tensor $g_{\nabla,\phi,T,\mathrm{Id},X}$. We use the gradient Ricci almost soliton equation to see

$$(\mathrm{Hess}_f + \rho - \lambda g)(\partial_{y_i}, \partial_{y_j}) = \partial^2_{y_i y_j} f = 0 \quad \text{for} \quad 1 \le i, j \le 2. \qquad (11.6.g)$$

Consequently, f is at most linear in y. This shows that there is a vector field Z and a smooth function h on M so that the potential function f has the form

$$f = \iota Z + \pi^* h. \qquad (11.6.h)$$

Our first task is to show that the auxiliary vector field $X = 0$ in defining the metric g. Suppose to the contrary that $X \neq 0$ at some point; we argue for a contradiction. Choose the system of local coordinates on M so that $X = \partial_{x^1}$. Expand $T = T_i^{\ j} dx^i \otimes \partial_{x^j}$ and $Z = Z^\ell \partial_{x^\ell}$. By Lemma 11.50, the soliton function $\lambda = \frac{1}{4}\tau$. Consequently,

$$\begin{aligned} 0 &= (\mathrm{Hess}_f + \rho - \tfrac{1}{4}\tau g)(\partial_{x^1}, \partial_{y_1}) \\ &= \tfrac{1}{4}\{2(T_1^1 - T_2^2 + 2(y_1 + \partial_{x^1} Z^1) - 2Z^1(3y_1^2 + 2y_1 T_1^1 + y_2 T_1^2 - 2\Gamma_{11}{}^1)\} \\ &\quad + \tfrac{1}{4}\{Z^2(4y_1 y_2 + 2y_1 T_2^1 + y_2(T_1^1 + T_2^2) - 4\Gamma_{12}{}^1)\}. \end{aligned}$$

All of the coefficients of this polynomial in (y_1, y_2) must vanish. This implies that $Z = 0$ so

$$0 = (\mathrm{Hess}_f + \rho - \tfrac{1}{4}\tau g)(\partial_{x^1}, \partial_{y_1}) = y_1 + \tfrac{1}{2}(T_1^1 - T_2^2).$$

This is not possible. Consequently, we conclude that $X = 0$ and as desired $g = g_{\nabla,\phi,T,\mathrm{Id}}$.

We now suppose that $\lambda \neq 0$. We must show that the vector field Z of Equation (11.6.h) vanishes. Suppose to the contrary that there is a point where Z is non-zero. We argue for a contradiction. Choose local coordinates so $Z = \partial_{x^1}$ and, consequently, $f = y_1 + \pi^* h$. We use the gradient Ricci almost soliton equation to see

$$\begin{aligned} 0 &= (\mathrm{Hess}_f + \rho - \tfrac{1}{4}\tau g)(\partial_{x^1}, \partial_{y_1}) = \tfrac{1}{2}\{(1 - 2y_1)T_1^1 - y_2 T_1^2 - T_2^2\} + \Gamma_{11}{}^1, \\ 0 &= (\mathrm{Hess}_f + \rho - \tfrac{1}{4}\tau g)(\partial_{x^2}, \partial_{y_2}) = \tfrac{1}{2}\{(4 - 2y_1)T_2^2 - y_2 T_1^2 - (1 + y_1)T_1^1\} + \Gamma_{12}{}^2. \end{aligned}$$

This shows that $T_1^1 = T_1^2 = T_2^2 = 0$. Since the scalar curvature of the metric $g_{\nabla,\phi,T,\mathrm{Id}}$ is given by $\tau = 3(T_1^1 + T_2^2) = 3\,\mathrm{Tr}\{T\}$, we have that $\tau = 0$. This implies that $\lambda = 0$ which is false. Consequently, $Z = 0$ so $f = \pi^* h$ and $\lambda = \frac{3}{4}\mathrm{Tr}\{T\}$. We use the gradient Ricci almost soliton equation to see

$$\begin{aligned} &(\mathrm{Hess}_f + \rho - \tfrac{1}{4}\tau g)(\partial_{x^1}, \partial_{y_1}) = \tfrac{1}{2}(T_1^1 - T_2^2) = 0, \\ &(\mathrm{Hess}_f + \rho - \tfrac{1}{4}\tau g)(\partial_{x^1}, \partial_{y_2}) = T_1^2 = 0, \quad \text{and} \quad (\mathrm{Hess}_f + \rho - \tfrac{1}{4}\tau g)(\partial_{x^2}, \partial_{y_1}) = T_2^1 = 0. \end{aligned}$$

This implies that $T = \psi(x^1, x^2)\,\mathrm{Id}$ is a multiple of the identity. for some smooth function ϕ on M. We compute:

$$\begin{aligned} (\mathrm{Hess}_f + \rho - \tfrac{1}{4}\tau g)(\partial_{x^1}, \partial_{x^1}) &= \tfrac{2}{3}y_1(\psi \partial_{x^1} h + \partial_{x^1}\psi) + \cdots, \\ (\mathrm{Hess}_f + \rho - \tfrac{1}{4}\tau g)(\partial_{x^2}, \partial_{x^2}) &= \tfrac{2}{3}y_1(\psi \partial_{x^2} h + \partial_{x^2}\psi) + \cdots. \end{aligned}$$

This shows that $\psi = Ce^{-h}$ so $T = Ce^{-h}\,\mathrm{Id}$ and $\lambda = \frac{3}{2}Ce^{-\pi^*h}$. We complete the proof by computing for $1 \le i, j \le 2$ that:

$$0 = (\mathrm{Hess}_f^{\mathcal{N}} + \rho - \tfrac{1}{4}\tau g)(\partial_{x^i}, \partial_{x^j}) = \tfrac{C}{2}e^{-h}\phi_{ij} - (\mathrm{Hess}_h^{\mathcal{M}} + 2\rho_{\nabla,s})(\partial_{x^i}, \partial_{x^j}). \qquad \square$$

Remark 11.53 Lemma 11.52 provides a method to construct steady traceless κ-Einstein solitons. Let $\mathcal{M} = (M, \nabla)$ be an affine surface and let $h \in C^\infty(M)$. Set

$$T := Ce^{-h}\,\mathrm{Id} \quad \text{and} \quad \phi := \tfrac{2}{C}e^h(\mathrm{Hess}_h + 2\rho_{s,\nabla}).$$

Then the modified Riemannian extension $g_{\nabla,\phi,T,\mathrm{Id}}$ of Equation (11.6.a) is a neutral signature metric on T^*M with scalar curvature $\tau = 6Ce^{-\pi^*h}$ and $(T^*M, g_{\nabla,\phi,T,\mathrm{Id}}, f = \pi^*h)$ is a gradient Ricci almost soliton with $\lambda = \frac{1}{4}\tau$. If h is a constant function, then $\mathrm{Hess}_h = 0$ and the modified Riemannian extension is Einstein; this fact was observed previously by Calviño-Louzao et al. [34, Theorem 2.1].

11.6.5 PRELIMINARY OBSERVATIONS IN THE ISOTROPIC CASE.

We assume henceforth that the gradient Ricci soliton is non-trivial. We proceed as in the proof of Lemma 11.50. Since $\nabla f \neq 0$ but $g(\nabla f, \nabla f) = 0$, we may find a local pseudo-orthonormal frame $\mathcal{B} = \{\nabla f, U, V, T\}$. We use Lemma 11.49 to see that for any vector fields X and Y,

$$W(\nabla f, V, X, Y) = W(U, T, X, Y), \quad W(U, V, X, Y) = 0, \quad W(\nabla f, T, X, Y) = 0. \qquad (11.6.i)$$

Lemma 11.54 Let $\mathcal{N} = (N, g, f)$ be an isotropic self-dual gradient Ricci soliton. Then $\mathcal{N}$ is steady, $\tau = 0$, $\mathrm{Ric}(\nabla f) = 0$, and the Ricci operator is 2-step nilpotent.

Proof. We use Lemma 11.5 to see that $2\,\mathrm{Ric}(\nabla f) = \nabla\tau = 2\lambda\nabla f$. This shows that ∇f is an eigenvector of the Ricci operator with eigenvalue λ. We use Equation (11.6.i) to see that

$$W(\nabla f, V, Z, \nabla f) = W(U, T, Z, \nabla f) \quad \text{and} \quad \mathfrak{C}(\nabla f, V, Z) = \mathfrak{C}(U, T, Z).$$

Consequently, Equation (11.6.b) implies that

$$0 = W(\nabla f, V, Z, \nabla f) - W(U, T, Z, \nabla f) = \left(\tfrac{\tau}{6} - \tfrac{2}{3}\lambda\right) g(\nabla f, Z).$$

Consequently, $\tau = 4\lambda$ is constant and $0 = 2\,\mathrm{Ric}(\nabla f) = 2\lambda f$. This shows that $\lambda = \tau = 0$ and the gradient Ricci soliton is steady. By Equation (11.6.f), $\mathrm{Ric}^2 = 0$. $\square$

Lemma 11.55 Let $\mathcal{N} = (N, g, f)$ be a 4-dimensional isotropic non-trivial gradient Ricci soliton. Then (N, g) is locally isometric to $(T^*M, g_{\nabla,\phi})$.

Proof. By Lemma 11.52, (N, g) is locally isometric to $(T^*M g_{\nabla,\phi,T,\mathrm{Id}})$. Let Γ be the Christoffel symbols of ∇; we will never deal with the Christoffel symbols of the Levi–Civita connection of g directly. Because $\tau = 3\,\mathrm{Tr}\{T\}$, we have that T is traceless (i.e., $T_1^1 = -T_2^2$). We also showed that $f = \iota Z + \pi^* h$ for some vector field Z on M and $h \in C^\infty(M)$. First suppose that $Z \neq 0$. Choose local coordinates (x^1, x^2) on M so that $Z = \partial_{x^1}$ and, consequently, that $f = y_1 + \pi^* h$. By Lemma 11.54, $\lambda = 0$. We may use the gradient Ricci soliton equation show that $T = 0$ and, therefore, $g = g_{\nabla,\phi}$ by computing:

$$\begin{aligned}
0 &= (\mathrm{Hess}_f + \rho)(\partial_{x^1}, \partial_{y_2}) = T_1^2 - \tfrac{1}{2}\left\{y_1 T_1^2 + \Gamma_{11}{}^2\right\}, \\
0 &= (\mathrm{Hess}_f + \rho)(\partial_{x^2}, \partial_{y_1}) = T_2^1 - \tfrac{1}{2}\{y_1 T_2^1 + \Gamma_{12}{}^1\}, \\
0 &= (\mathrm{Hess}_f + \rho)(\partial_{x^1}, \partial_{y_1}) = -T_2^2 - \tfrac{1}{2}\{y_2 T_1^2 + 2y_1 T_2^2 - \Gamma_{11}{}^1\}.
\end{aligned}$$

On the other hand, if $Z = 0$, then $f = \pi^* h$. This shows that $(\mathrm{Hess}_f + \rho)(\partial_{x^1}, \partial_{y_2}) = T_1^2$, $(\mathrm{Hess}_f + \rho)(\partial_{x^2}, \partial_{y_1}) = T_2^1$, and $(\mathrm{Hess}_f + \rho)(\partial_{x^1}, \partial_{y_1}) = -T_2^2$. Consequently, we once again have that $T = 0$ and $g = g_{\nabla,\phi}$. □

The underlying structure of any self-dual isotropic gradient Ricci soliton is that of a deformed Riemannian extension. We will now analyze the existence of gradient Ricci solitons on the cotangent bundle T^*M of an affine surface (M, ∇) equipped with the metric $g_{\nabla,\phi}$ of Equation (11.6.a).

Lemma 11.56 Any non-trivial gradient Ricci soliton $(T^*M, g_{\nabla,\phi}, f)$ is steady and $f = \pi^* h$.

Proof. By Equation (11.6.g), the potential function of any Ricci soliton on $(T^*M, g_{\nabla,\phi})$ has the form $f = \iota Z + \pi^* h$ for some vector field Z on M and some $h \in C^\infty(M)$. Suppose that Z does not vanish identically. We argue for a contradiction. Choose local coordinates (x^1, x^2) on M so that $Z = \partial_{x^1}$ to express $f = y_1 + \pi^* h$. We then have $\mathrm{Hess}_f(\partial_{x^i}, \partial_{y_j}) = \Gamma_{1i}{}^j$. By Lemma 11.40 and the Ricci soliton equation, we have that

$$\Gamma_{11}{}^1 = \Gamma_{12}{}^2 = \lambda \quad \text{and} \quad \Gamma_{11}{}^2 = \Gamma_{12}{}^1 = 0\,. \tag{11.6.j}$$

Thus, by Lemma 11.40, the only possibly non-zero components of the Ricci tensor of ρ_g are

$$\rho_g(\partial_{x^1}, \partial_{x^2}) = -2\partial_{x^1}\Gamma_{22}{}^2 \quad \text{and} \quad \rho_g(\partial_{x^2}, \partial_{x^2}) = 2\partial_{x^1}\Gamma_{22}{}^1\,. \tag{11.6.k}$$

By Equation (11.6.j), $\|\nabla f\|^2 = 2\lambda\, y_1 + 2\partial_{x^1} h(x^1, x^2) - \phi_{11}(x^1, x^2)$. Consequently, if $\lambda \neq 0$, then $\|\nabla f\|^2 \neq 0$ on an open dense subset of T^*M so, by Theorem 11.45, $(T^*M, g_{\nabla,\phi})$ is locally conformally flat. We compute that $0 = W(\partial_{x^1}, \partial_{y_1}, \partial_{x^2}, \partial_{x^1}) = \frac{1}{2}\partial_{x^1}\Gamma_{22}{}^2$ and, therefore,

$\Gamma_{22}{}^2(x^1, x^2) = \Gamma_{22}{}^2(x^2)$. We use the Ricci soliton equation to see

$$\begin{aligned} 0 &= \operatorname{Hess}_f(\partial_{x^2}, \partial_{x^2}) + \rho_g(\partial_{x^2}, \partial_{x^2}) - \lambda g_{\nabla,\phi}(\partial_{x^2}, \partial_{x^2}) \\ &= \partial_{x^2}\partial_{x^2}h + \Gamma_{22}{}^2(\phi_{12} - \partial_{x^2}h) - \lambda\phi_{22} - \partial_{x^2}\phi_{12} + \Gamma_{22}{}^1(\phi_{11} - \partial_{x^1}h) \\ &\quad + \tfrac{1}{2}\partial_{x^1}\phi_{22} + (2 - y_1)\partial_{x^1}\Gamma_{22}{}^1 . \end{aligned}$$

We examine the coefficient of y_1 to see that $\partial_{x^1}\Gamma_{22}{}^1 = 0$. Now, it follows from Equation (11.6.k) that $(T^*M, g_{\nabla,\phi})$ is Ricci flat. This shows that any gradient Ricci soliton on $(T^*M, g_{\nabla,\phi})$ with potential function $f = \iota Z + \pi^*h$ is trivial if $Z \neq 0$. We may therefore assume $Z = 0$ so $f = \pi^*h$. We show that $\lambda = 0$ and complete the proof by computing:

$$\operatorname{Hess}_f(\partial_{x^1}, \partial_{y_1}) + \rho(\partial_{x^1}, \partial_{y_1}) - \lambda g_{\nabla,\phi}(\partial_{x^1}, \partial_{y_1}) = -\lambda = 0 . \qquad \square$$

11.6.6 THE PROOF OF LEMMA 11.42 AND OF THEOREM 11.47. Adopt the notation of Equation (11.6.a) to define $g_{\nabla,\phi}$. Suppose that $(T^*M, g_{\nabla,\phi}, f)$ is a gradient Ricci soliton. We apply Lemma 11.56 to see that $f = \pi^*h$. The possibly non-vanishing terms of Hess_f are given by $\operatorname{Hess}_f(\partial_{x^i}, \partial_{x^j}) = \operatorname{Hess}_h(\partial_{x^i}, \partial_{x^j})$. Since the soliton is steady by Lemma 11.56, we may use Lemma 11.40 to conclude that $\operatorname{Hess}_f + \rho = 0$ if and only if $\operatorname{Hess}_h + 2\rho_{\nabla,s} = 0$. $\square$

Remark 11.57

1. We note that the existence of a gradient Ricci soliton on $(T^*M, g_{\nabla,\phi})$ is independent of the choice of the symmetric $(0,2)$-tensor field ϕ. Hence, any affine gradient Ricci soliton (M, ∇, h) induces an infinite family of steady gradient Ricci solitons $(T^*M, g_{\nabla,\phi}, \pi^*h)$. All such gradient Ricci solitons are isotropic since $\|\nabla\pi^*h\|^2 = 0$.
2. The conformal structure of $(T^*M, g_{\nabla,\phi})$ is related to the projective structure of the affine surface (M, ∇). The possibly non-zero components of the Weyl tensor are, up to the usual $\mathbb{Z}_2$-symmetries, given by:

$$\begin{aligned} W(\partial_{x^1}, \partial_{x^2}, \partial_{x^1}, \partial_{y_1}) &= W(\partial_{x^1}, \partial_{x^2}, \partial_{x^2}, \partial_{y_2}) = \frac{1}{2}\{\rho^D(\partial_{x^1}, \partial_{x^2}) - \rho^D(\partial_{x^2}, \partial_{x^1})\}, \\ W(\partial_{x^1}, \partial_{x^2}, \partial_{x^1}, \partial_{x^2}) &= \Theta(\phi) + y_1\{(D_{\partial_{x^1}}\rho^D)(\partial_{x^2}, \partial_{x^2}) - (D_{\partial_{x^2}}\rho^D)(\partial_{x^2}, \partial_{x^1})\} \\ &\quad + y_2\{(D_{\partial_{x^2}}\rho^D)(\partial_{x^1}, \partial_{x^1}) - (D_{\partial_{x^1}}\rho^D)(\partial_{x^1}, \partial_{x^2})\} . \end{aligned}$$

   Here, Θ is a polynomial in the derivatives of the components ϕ_{ij} up to order two. Consequently, if $(T^*M, g_{\nabla,\phi})$ is locally conformally flat, then ρ_∇ and $\nabla\rho_\nabla$ are symmetric, i.e., the affine connection ∇ is projectively flat with symmetric Ricci tensor.

11.6.7 INHOMOGENEOUS AFFINE GRADIENT RICCI SOLITONS. We now describe a simple ansatz for obtaining affine surfaces which admit affine gradient Ricci solitons. We do not sum over repeated indices in this section. If f is a smooth function, let $f_i := \partial_{x^i} f$, $f_{ij} := \partial_{x^j} f_i$, and so forth. Consider the following partial differential equation:

$$f_{12} = (f_{112} f_1 - f_{11} f_{12}) f_1^{-2} + (f_{122} f_2 - f_{22} f_{12}) f_2^{-2} . \tag{11.6.1}$$

Theorem 11.58 *Suppose f is a solution to Equation (11.6.1) on open set $\mathcal{O} \subset \mathbb{R}^2$. Let ∇ be the affine connection with non-zero Christoffel symbols*

$$\Gamma_{11}{}^1 := f_{11} f_1^{-1} \quad \text{and} \quad \Gamma_{22}{}^2 := f_{22} f_2^{-1} .$$

Then $\mathcal{M} := (\mathcal{O}, \nabla, f)$ is an affine gradient Ricci soliton.

Proof. We use Lemma 11.40 to compute the Ricci tensor:

$$\begin{aligned}
\rho_{jk} &= \partial_{x^i} \Gamma_{jk}{}^i - \partial_{x^j} \Gamma_{ik}{}^i + \Gamma_{in}{}^i \Gamma_{jk}{}^n - \Gamma_{jn}{}^i \Gamma_{ik}{}^n \\
&= \delta_{jk} \partial_{x^j} \Gamma_{jj}{}^j - \partial_{x^j} \Gamma_{kk}{}^k + \delta_{jk} \{\Gamma_{jj}{}^j \Gamma_{jj}{}^j - \Gamma_{jj}{}^j \Gamma_{jj}{}^j\} \\
&= \left\{ \begin{array}{ll} 0 & \text{if} \quad j = k \\ -\partial_{x^j} \Gamma_{kk}{}^k & \text{if} \quad j \neq k \end{array} \right\},
\end{aligned}$$

$$\rho_s(\partial_{x^i}, \partial_{x^j}) = \left\{ \begin{array}{ll} 0 & \text{if} \quad i = j \\ -\frac{1}{2}\{\partial_j \Gamma_{ii}{}^i + \partial_i \Gamma_{jj}{}^j\} & \text{if} \quad i \neq j \end{array} \right\},$$

$$H_f(\partial_{x^i}, \partial_{x^j}) = \left\{ \begin{array}{ll} f_{ii} - f_i \Gamma_{ii}{}^i & \text{if} \quad i = j \\ f_{ij} & \text{if} \quad i \neq j \end{array} \right\}.$$

Since $f_{ii} - f_i \Gamma_{ii}{}^i = 0$, $\text{Hess}_{h,ii} + 2\rho_{s,ii} = \text{Hess}_{f,ii} = 0$. We use Equation (11.6.1) to see that $\text{Hess}_{f,12} + 2\rho_{s,12} = 0$. Consequently, $(\mathcal{O}, \nabla, f)$ is an affine gradient Ricci soliton. □

We can rewrite Equation (11.6.1) in the form $\partial_{x^1} \partial_{x^2} \{f - \log(f_1) - \log(f_2)\} = 0$. In the special case that $f(x^1, x^2) = \xi(x^1 + x^2)$, we obtain the ODE $\{\xi - 2\log(\xi')\}'' = 0$. This implies that $\xi - 2\log(\xi') = a + bt$ and, consequently,

$$\xi(t) = \left\{ \begin{array}{ll} -2\log(\frac{1}{2b}(e^{-a-bt} + c) & \text{if}\, b \neq 0 \\ -2\log(\frac{1}{2}(e^{-a/2} t + c)) & \text{if}\, b = 0. \end{array} \right\} .$$

If $b \neq 0$, then $\|\nabla \rho\|_\rho^2 = -\frac{2}{bc} e^{-\frac{a}{2} - \frac{b}{2} t}$ so this geometry is not homogeneous. If $b = 0$, this is a Type $\mathcal{B}$ geometry (which is homogeneous) since $\Gamma_{11}{}^1 = \Gamma_{22}{}^2 = -(x^1 + x^2 + e^{a/2} c)^{-1}$. If we take $ds^2 = \eta(x^1 + x^2)(dx^1 \otimes dx^2 + dx^2 \otimes dx^1)$, then the only non-zero Christoffel symbols are $\Gamma_{11}{}^1 = \Gamma_{22}{}^2 = (\eta_1 \eta)(x^1 + x^2)$. Consequently, by choosing η appropriately, we see that all these geometries arise as the Levi–Civita connection of a recurrent hyperbolic Einstein metric.

We return to the general setting. Let $\mathcal{F} := (f, f_1, f_2, f_{11}, f_{12}, f_{12})$. We can express Equation (11.6.l) in the form

$$\begin{aligned} &a_{112}(\mathcal{F}) f_{112} + a_{122}(\mathcal{F}) f_{122} + a(\mathcal{F}) f_{112} = 0 \quad \text{where} \quad a_{112}(\mathcal{F}) = f_1^{-1}, \\ &a_{122} = f_2^{-1}, \quad \text{and} \quad a(\mathcal{F}) = f_{11} f_{12} f_1^{-2} + f_{22} f_{12} f_2^{-1} - f_{12} . \end{aligned} \tag{11.6.m}$$

For $\delta \in \mathbb{R}$, make a change of variables setting $\tilde{x}^1 = x^1 + \delta x^2$ and $\tilde{x}^2 = x^2$. This converts Equation (11.6.m) into

$$\tilde{a}_{111}(\mathcal{F}) \tilde{f}_{111} + \tilde{a}_{112}(\mathcal{F}) \tilde{f}_{112} + \tilde{a}_{122}(\mathcal{F}) \tilde{f}_{122} + \tilde{a}_{222}(\mathcal{F}) \tilde{f}_{222} + \tilde{a}(\mathcal{F}) = 0 \tag{11.6.n}$$

where $\tilde{a}_{111}(\mathcal{F}) = \delta f_1^{-1} + \delta^2 f_2^{-1}$. We work on a small neighborhood of 0 and assume $f_1(0) \neq 0$ and $f_2(0) \neq 0$. For generic choice of δ, $\tilde{a}_{111}$ will be non-zero. Consequently, we can use the *Cauchy–Kovalevskaya Theorem* (see, for example, Evans [63]) to solve this partial differential equation with $\tilde{f}(0, \tilde{x}^2) = \eta_0(x^2)$, $\tilde{f}_1(0, x^2) = \eta_1(x^2)$, and $\tilde{f}_{11}(0, x^2) = \eta_2(\tilde{x}^2)$ given arbitrarily. In particular, we can specify the 3-jets of $\tilde{f}$ (or equivalently of f) at the origin arbitrarily subject to the single condition imposed by Equation (11.6.n). Let $\mathcal{B} = J^3(f)(0)$ be the collection of 3-jets of f at the origin:

$$\mathcal{B} = \{f(0), f_1(0), f_2(0), f_{11}(0), f_{12}(0), f_{22}(0), f_{111}(0), f_{112}(0), f_{122}(0), f_{222}(0)\} .$$

Suppose given constants ϱ_{12} and ϱ_{21}. We impose the relations

$$\begin{aligned} &b_1 \neq 0, \; b_2 \neq 0, \; b_{12} + 2(\varrho_{12} + \varrho_{21}) = 0, \\ &\varrho_{12} = (b_{112} b_1 - b_{11} b_{12}) b_1^{-2}, \; \varrho_{21} = (b_{122} b_2 - b_{22} b_{12}) b_2^{-2} . \end{aligned}$$

For example, we could take

$$\begin{aligned} &b_1 = b_2 = 1, \quad b_{11} = b_{22} = 0, \quad b_{112} = \varrho_{12}, \\ &b_{122} = \varrho_{21}, \quad b_{111} = b_{222} = 0, \quad b_{12} = -2(\varrho_{12} + \varrho_{21}) . \end{aligned}$$

We would then have $\rho_{12}(0) = \varrho_{12}$, $\rho_{21}(0) = \varrho_{21}$, and f would satisfy Equation (11.6.l) at the origin. Consequently, as noted above, we can find a solution to Equation (11.6.l) with $\rho_{12}(0) = \varrho_{12}$ and $\rho_{21}(0) = \varrho_{21}$ given arbitrarily. Thus, in particular, the Ricci tensor need not be symmetric. Therefore, the affine structure is not the Levi–Civita connection of any metric. By examining the 5-jets (and normalizing them using the relations imposed by differentiating Equation (11.6.l)), and by applying the argument given in the case where $f = \xi(x^1 + x^2)$, we see that f is generically inhomogeneous as well. We omit the details in the interests of brevity.

A. Cauchy (1789–1857)

S. Kovalevskaya (1850–1891)

Bibliography

[1] Z. Afifi, Riemann extensions of affine connected spaces, *Quart. J. Math., Oxford Ser. (2)* **5** (1954), 312–320. DOI: 10.1093/qmath/5.1.312.

[2] A. Alcolado, A. MacDougall, A. Coley, and S. Hervik, 4D neutral signature VSI and CSI spaces, *J. Geom. Phys.* **62** (2012), 594–603. DOI: 10.1016/j.geomphys.2011.04.012.

[3] D. Alekseevski, Self-similar Lorentzian manifolds, *Ann. Global Anal. Geom.* **3** (1985), 59–84. DOI: 10.1007/bf00054491.

[4] D. Alekseevsky, V. Cortés, A. S. Galaev, and T. Leistner, Cones over pseudo-Riemannian manifolds and their holonomy, *J. Reine Angew. Math.* **635** (2009), 23–69. DOI: 10.1515/crelle.2009.075.

[5] C. Allendoerfer and A. Weil, The Gauss–Bonnet theorem for Riemannian polyhedra, *Trans. Amer. Math. Soc.* **53** (1943), 101–129. DOI: 10.1090/s0002-9947-1943-0007627-9.

[6] L. J. Alty, The generalized Gauss–Bonnet–Chern theorem, *J. Math. Phys.* **36** (1995), 3094–3105. DOI: 10.1063/1.531015.

[7] T. Arias-Marco and O. Kowalski, Classification of locally homogeneous affine connections with arbitrary torsion on 2-dimensional manifolds, *Monatsh. Math.* **153** (2008), 1–18. DOI: 10.1007/s00605-007-0494-0.

[8] M. F. Atiyah, R. Bott, and V. K. Patodi, On the heat equation and the index theorem, *Invent. Math.* **19** (1973), 279–330; (Errata **28** (1975), 277–280). DOI: 10.1007/bf01425417.

[9] A. Avez, Formule de Gauss–Bonnet–Chern en métrique de signature quelconque, *C. R. Acad. Sci. Paris* **256** (1962), 2049–2051.

[10] W. Batat, M. Brozos-Vázquez, E. García-Río, and S. Gavino-Fernández, Ricci solitons on Lorentzian manifolds with large isometry groups, *Bull. Lond. Math. Soc.* **43** (2011), 1219–1227. DOI: 10.1112/blms/bdr057.

[11] L. Berard Bergery and A. Ikemakhen, On the holonomy of Lorentzian manifolds, *Proc. Sympos. Pure Math* **54** (1993), 27–40. DOI: 10.1090/pspum/054.2/1216527.

[12] M. Berger, Quelques formules de variation pour une structure riemannienne, *Ann. Scient. Éc. Norm. Supér.* **3** (1970), 285–294. DOI: 10.24033/asens.1194.

[13] A. L. Besse, *Einstein manifolds*, Classics in Mathematics, Springer-Verlag, Berlin, 2008. DOI: 10.1007/978-3-540-74311-8.

[14] E. Boeckx, O. Kowalski, and L. Vanhecke, *Riemannian manifolds of conullity two*, World Scientific Publishing Co., Inc., River Edge, NJ, 1996. DOI: 10.1142/9789812819970.

[15] M. Brozos-Vázquez, G. Calvaruso, E. García-Río, and S. Gavino-Fernández, Three-dimensional Lorentzian homogeneous Ricci solitons, *Israel J. Math.* **188** (2012), 385–403. DOI: 10.1007/s11856-011-0124-3.

[16] M. Brozos-Vázquez and E. García-Río, Four-dimensional neutral signature self-dual gradient Ricci solitons, *Indiana Univ. Math. J.* **65** (2016), 1921–1943. DOI: 10.1512/iumj.2016.65.5938.

[17] M. Brozos-Vázquez, E. García-Río, and S. Gavino-Fernández, Locally conformally flat Lorentzian gradient Ricci solitons, *J. Geom. Anal.* **23** (2013), 1196–1212. DOI: 10.1007/s12220-011-9283-z.

[18] M. Brozos-Vázquez, E. García-Río, S. Gavino-Fernández, and P. Gilkey, The structure of the Ricci tensor on locally homogeneous Lorentzian gradient Ricci solitons, arXiv:1403.4400, to appear in *Proc. Roy. Soc. Edinburgh Sect. A.*

[19] M. Brozos-Vázquez, E. García-Río, and P. Gilkey, Homogeneous affine surfaces: moduli spaces, *J. Math. Anal. Appl.* **444** (2016), 1155–1184. DOI: 10.1016/j.jmaa.2016.07.005.

[20] M. Brozos-Vázquez, E. García-Río, and P. Gilkey, Homogeneous affine surfaces: affine Killing vector fields and gradient Ricci solitons, arXiv:1512.05515, to appear in *J. Math. Soc. Japan.*

[21] M. Brozos-Vázquez, E. García-Río, P. Gilkey, S. Nikčević, and R. Vázquez-Lorenzo, *The geometry of Walker manifolds*, Synthesis Lectures on Mathematics and Statistics **5**, Morgan & Claypool Publishers, Williston, VT, 2009. DOI: 10.2200/s00197ed1v01y200906mas005.

[22] M. Brozos-Vázquez, E. García-Río, P. Gilkey, and X. Valle-Regueiro, Half conformally flat generalized quasi-Einstein manifolds, arXiv:1702.06714 [math.DG].

[23] M. Brozos-Vázquez, E. García-Río, and X. Valle-Regueiro, Conformally flat Lorentzian generalized quasi-Einstein metrics, *in preparation*.

[24] M. Brozos-Vázquez, E. García-Río, and R. Vázquez-Lorenzo, Complete locally conformally flat manifolds of negative curvature, *Pacific J. Math.* **226** (2006), 201–219. DOI: 10.2140/pjm.2006.226.201.

[25] M. Brozos-Vázquez, E. García-Río, and R. Vázquez-Lorenzo, Osserman and conformally Osserman manifolds with warped and twisted product structure, *Results Math.* **52** (2008), no. 3-4, 211–221. DOI: 10.1007/s00025-008-0306-4.

[26] M. Brozos-Vázquez, P. Gilkey, and S. Nikčević, *Geometric Realizations of Curvature*, ICP Advanced Texts in Mathematics V **6**, Imperial College Press, London United Kingdom, 2012. DOI: 10.1142/9781848167421.

[27] R. Bryant, Ricci flow solitons in dimension three with SO(3)-symmetries, available at `www.math.duke.edu/~bryant/3DRotSymRicciSolitons.pdf`

[28] M. Cahen, J. Leroy, M. Parker, F. Tricerri, and L. Vanhecke, Lorentz manifolds modelled on a Lorentz symmetric space, *J. Geom. Phys.* **7** (1990), 571–581. DOI: 10.1016/0393-0440(90)90007-p.

[29] M. Cahen and N. Wallach, Lorentzian symmetric spaces, *Bull. Amer. Math. Soc.* **76** (1970), 585–591. DOI: 10.1090/s0002-9904-1970-12448-x.

[30] G. Calvaruso, Homogeneous structures on three-dimensional Lorentzian manifolds, *J. Geom. Phys.* **57** (2007), 1279–1291. DOI: 10.1016/j.geomphys.2006.10.005.

[31] G. Calvaruso and A. Fino, Ricci solitons and geometry of four-dimensional non-reductive homogeneous spaces, *Canad. J. Math.* **64** (2012), 778–804. DOI: 10.4153/cjm-2011-091-1.

[32] G. Calvaruso and O. Kowalski, On the Ricci operator of locally homogeneous Lorentzian 3-manifolds, *Cent. Eur. J. Math.* **7** (2009), 124–139. DOI: 10.2478/s11533-008-0061-5.

[33] E. Calviño-Louzao, M. Fernández-López, E. García-Río, and R. Vázquez-Lorenzo, Homogeneous Ricci almost solitons, *Israel J. Math.*, to appear.

[34] E. Calviño-Louzao, E. García-Río, P. Gilkey, and R. Vázquez-Lorenzo, The geometry of modified Riemannian extensions, *Proc. Roy. Soc. Lond. Ser. A Math. Phys. Eng. Sci.* **465** (2009), 2023–2040. DOI: 10.1098/rspa.2009.0046.

[35] E. Calviño-Louzao, E. García-Río, M. E. Vázquez-Abal, and R. Vázquez-Lorenzo, Curvature operators and generalizations of symmetric spaces in Lorentzian geometry, *Adv. Geom.* **12** (2012), 83–100. DOI: 10.1515/advgeom.2011.036.

[36] A. M. Candela, J. L. Flores, and M. Sánchez, On general plane fronted waves. Geodesics, *Gen. Rel. Gravit.*, **35** (2003), 631–649. DOI: 10.1023/A:1022962017685.

[37] H.-D. Cao, Geometry of complete gradient shrinking Ricci solitons, *Geometry and analysis. No. 1*, 227–246, *Adv. Lect. Math. (ALM)* **17**, Int. Press, Somerville, MA, 2011.

[38] H.-D. Cao and Q. Chen, On locally conformally flat steady gradient Ricci solitons, *Trans. Amer. Math. Soc* **364** (2012), 2377–2391. DOI: 10.1090/s0002-9947-2011-05446-2.

[39] X. Cao, B. Wang, and Z. Zhang, On locally conformally flat gradient shrinking Ricci solitons, *Commun. Contemp. Math.* **13** (2011), 269–282. DOI: 10.1142/s0219199711004191.

[40] G. Catino, L. Cremaschi, Z. Djadli, C. Mantegazza, and L. Mazzieri, The Ricci-Bourguignon flow, arXiv:1507.00324, to appear in *Pacific J. Math.* DOI: 10.2140/pjm.2017.287.337.

[41] L. F. di Cerbo, Generic properties of homogeneous Ricci solitons, *Adv. Geom.* **14** (2014), 225–237. DOI: 10.1515/advgeom-2013-0031.

[42] B.-L. Chen, Strong uniqueness of the Ricci flow, *J. Diff. Geom.* **82** (2009), 363–382. DOI: 10.4310/jdg/1246888488.

[43] X. Chen and Y. Wang, On four-dimensional anti-self-dual gradient Ricci solitons, *J. Geom. Anal.* **25** (2015), 1335–1343. DOI: 10.1007/s12220-014-9471-8.

[44] S. Chern, A simple intrinsic proof of the Gauss–Bonnet formula for closed Riemannian manifolds, *Ann. Math.* **45** (1944), 747–752. DOI: 10.2307/1969302.

[45] S. Chern, Characteristic classes of Hermitian manifolds, *Ann. Math.* **47** (1946), 85–121. DOI: 10.2307/1969037.

[46] S. Chern, Pseudo-Riemannian geometry and the Gauss–Bonnet formula, *Ann. Acad. Brasil. Ci.* **35** (1963), 17–26.

[47] B. Chow, S.-Ch. Chu, D. Glickenstein, C. Guenther, J. Isenberg, T. Ivey, D. Knopf, P. Lu, F. Luo, and L. Ni, *The Ricci flow: techniques and applications. Part I. Geometric aspects*, Mathematical Surveys and Monographs, **135**. American Mathematical Society, Providence, RI, 2007. DOI: 10.1090/surv/135.

[48] A. Coley, S. Hervik, and N. Pelavas, Lorentzian spacetimes with constant curvature invariants in three dimensions, *Class. Quant. Grav.* **25** (2008), 025008, 14 pp. DOI: 10.1088/0264-9381/25/2/025008.

[49] A. Coley, S. Hervik, and N. Pelavas, Lorentzian manifolds and scalar curvature invariants, *Class. Quant. Grav.* **27** (2010), 102001, 9 pp. DOI: 10.1088/0264-9381/27/10/102001.

[50] A. Coley, R. Milson, V. Pravda, and A. Pravdová, Vanishing scalar invariant spacetimes in higher dimensions, *Class. Quant. Grav.* **21** (2004), 5519–5542. DOI: 10.1088/0264-9381/21/23/014.

[51] S. Console and L. Nicolodi, Infinitesimal characterization of almost Hermitian homogeneous spaces, *Comment. Math. Univ. Carolin.* **40** (1999), 713–721.

[52] A. Dancer and M. Y. Wang, On Ricci solitons of cohomogeneity one, *Ann. Global Anal. Geom.* **39** (2011), 259–292. DOI: 10.1007/s10455-010-9233-1.

[53] A. Derdzinski, Non-Walker self-dual neutral Einstein four-manifolds of Petrov type III, *J. Geom. Anal.* **19** (2009), 301–357. DOI: 10.1007/s12220-008-9066-3.

[54] A. Derdzinski, Ricci solitons, *Wiad. Mat.* **48** (2012), 1–32.

[55] A. Derdzinski and W. Roter, Projectively flat surfaces, null parallel distributions, and conformally symmetric manifolds, *Tohoku Math. J.* **59** (2007), 565–602. DOI: 10.2748/tmj/1199649875.

[56] J. C. Díaz-Ramos, E. García-Río, and L. Nicolodi, Curvature invariants and locally homogeneous metric G-structures, preprint.

[57] J. C. Díaz-Ramos, E. García-Río, and R. Vázquez-Lorenzo, Four-dimensional Osserman metrics with nondiagonalizable Jacobi operators, *J. Geom. Anal.* **16** (2006), 39–52. DOI: 10.1007/bf02930986.

[58] M. Dunajski, and S. West, Anti-self-dual conformal structures in neutral signature, *Recent developments in pseudo-Riemannian geometry*, 113–148, ESI Lect. Math. Phys., Eur. Math. Soc., Zürich, 2008. DOI: 10.4171/051-1/4.

[59] C. Dunn and C. McDonald, Singer invariants and various types of curvature homogeneity, *Ann Glob. Anal. Geom.* **45** (2014), 303–317. DOI: 10.1007/s10455-013-9403-z.

[60] M. Eminenti, G. La Nave, and C. Mantegazza, Ricci solitons: the equation point of view, *Manuscripta Math.* **127** (2008), 345–367. DOI: 10.1007/s00229-008-0210-y.

[61] Y. Euh, J. H. Park, and K. Sekigawa, A curvature identity on a 4-dimensional Riemannian manifold, *Results Math.* **63** (2013), 107–114. DOI: 10.1007/s00025-011-0164-3.

[62] Y. Euh, J. H. Park, and K. Sekigawa, A curvature identity on 6-dimensional Riemannian manifold and its applications, *Czechoslovak Math. J.* **67** (2017), 253–270. DOI: 10.21136/CMJ.2017.0540-15.

[63] L. Evans, *Partial differential equations*, Graduate Texts in Mathematics **19**, American Mathematical Society, Providence R. I. DOI: 10.1112/blms/20.4.375.

[64] M. Fernández-López and E. García-Río, A note on locally conformally flat gradient Ricci solitons, *Geom. Dedicata* **168** (2014), 1–7. DOI: 10.1007/s10711-012-9815-0.

[65] M. Fernández-López and E. García-Río, On gradient Ricci solitons with constant scalar curvature, *Proc. Amer. Math. Soc.* **144** (2016), 369–378. DOI: 10.1090/proc/12693.

[66] M. Fernández-López, E. García-Río, D. N. Kupeli, and B. Unal, A curvature condition for a twisted product to be a warped product, *Manuscripta Math.* **106** (2001), 213–217. DOI: 10.1007/s002290100204.

[67] E. García-Río, P. Gilkey, and S. Nikčević, Homogeneity of Lorentzian three-manifolds with recurrent curvature, *Math. Nachr.* **287** (2014), 32–47. DOI: 10.1002/mana.201200302.

[68] E. García-Río, P. Gilkey, and S. Nikčević, Homothety curvature homogeneity and homothety homogeneity, *Ann. Global Anal. Geom.* **48** (2015), 149–170. DOI: 10.1007/s10455-015-9462-4.

[69] E. García-Río, P. Gilkey, S. Nikčević, and R. Vázquez-Lorenzo, *Applications of affine and Weyl geometry*, Synthesis Lectures on Mathematics and Statistics **13**, Morgan & Claypool Publ., Williston, VT, 2013. DOI: 10.2200/s00502ed1v01y201305mas013.

[70] E. García-Río and D. N. Kupeli, Some splitting theorems for stably causal spacetimes, *Gen. Relativ. Gravit.* **30** (1998), 35–44. DOI: 10.1023/A:1018816815433.

[71] P. Gilkey, Curvature and the eigenvalues of the Laplacian for elliptic complexes, *Adv. Math.* **10** (1973), 344–382. DOI: 10.1016/0001-8708(73)90119-9.

[72] P. Gilkey, *Invariance theory, the heat equation, and the Atiyah-Singer index theorem*. Second edition. Studies in Advanced Mathematics. CRC Press, Boca Raton, FL, 1995.

[73] P. Gilkey, *The Geometry of curvature homogeneous pseudo-Riemannian manifolds*, ICP Advanced Texts in Mathematics, **2**. Imperial College Press, London, 2007. DOI: 10.1142/9781860948589.

[74] P. Gilkey and S. Nikčević, Curvature homogeneous spacelike Jordan Osserman pseudo-Riemannian manifolds, *Class. Quant. Grav.* **21** (2004), 497–507. DOI: 10.1088/0264-9381/21/2/013.

[75] P. Gilkey and S. Nikčević, Complete k-curvature homogeneous pseudo-Riemannian manifolds, *Ann. Global Anal. Geom.* **27** (2005), 87–100. DOI: 10.1007/s10455-005-5217-y.

[76] P. Gilkey and J. H. Park, Analytic continuation, the Chern–Gauss–Bonnet theorem, and the Euler–Lagrange equations in Lovelock theory for indefinite signature metrics, *J. Geom. Phys.* **88** (2015), 88–93. DOI: 10.1016/j.geomphys.2014.11.006.

[77] P. Gilkey, J. H. Park, and K. Sekigawa, Universal curvature identities, *Diff. Geom. Appl.* **29** (2011), 770–778. DOI: 10.1016/j.difgeo.2011.08.005.

[78] P. Gilkey, J. H. Park, and K. Sekigawa, Universal curvature identities and Euler Lagrange formulas for Kähler manifolds, *J. Math. Soc. Japan* **68** (2016), 459–487. DOI: 10.2969/jmsj/06820459.

[79] A. Haji-Badali, Ricci almost solitons on three-dimensional manifolds with recurrent curvature, *Mediterr. J. Math.* (2017) **14** (2017), no. 1, Art. 4, 9 pp. DOI: 10.1007/s00009-016-0810-9.

[80] S. Helgason, *Differential geometry, Lie groups and symmetric spaces*, Pure and Applied Mathematics **80**. Academic Press, Inc., New York-London, 1978. DOI: 10.1090/gsm/034.

[81] F. Hirzebruch, *Topological methods in algebraic geometry*, Classics in Mathematics. Springer-Verlag, Berlin, 1995. DOI: 10.1007/978-3-642-62018-8.

[82] Th. Ivey, New examples of complete Ricci solitons, *Proc. Amer. Math. Soc.* **122** (1994), 241–245. DOI: 10.1090/s0002-9939-1994-1207538-5.

[83] M. Jablonski, Homogeneous Ricci solitons are algebraic, *Geom. Topol.* **18** (2014), 2477–2486. DOI: 10.2140/gt.2014.18.2477.

[84] M. Kanai, On a differential equation characterizing a Riemannian structure of a manifold, *Tokyo J. Math.* **6** (1983), 143–151. DOI: 10.3836/tjm/1270214332.

[85] S. Kobayashi and K. Nomizu, *Foundations of differential geometry*. Vol. I, Interscience Publishers, a division of John Wiley & Sons, New York-London 1963.

[86] S. Kobayashi and K. Nomizu, *Foundations of differential geometry*. Vol. II, Interscience Publishers, a division of John Wiley & Sons, New York-London 1969.

[87] B. Kostant, Holonomy and the Lie algebra of infinitesimal motions of a Riemannian manifold, *Trans. Amer. Math. Soc.* **80** (1955), 528–542. DOI: 10.1007/b94535_1.

[88] B. Kotschwar, On rotationally invariant shrinking Ricci solitons, *Pacific J. Math.* **236** (2008), 73–88. DOI: 10.2140/pjm.2008.236.73.

[89] O. Kowalski and S. Nikčević, On Ricci eigenvalues of locally homogeneous Riemannian 3-manifolds, *Geom. Dedicata* **62** (1996), 65–72. DOI: 10.1007/bf00240002.

[90] O. Kowalski and F. Prüfer, On Riemannian 3-manifolds with distinct constant Ricci eigenvalues, *Math. Ann.* **300** (1994), 17–28. DOI: 10.1007/bf01450473.

[91] O. Kowalski and A. Vanžurová, On curvature-homogeneous spaces of type $(1, 3)$, *Math. Nachr.* **284** (2011), 2127–2132. DOI: 10.1002/mana.201000008.

[92] O. Kowalski and A. Vanžurová, On a generalization of curvature homogeneous spaces, *Results Math.* **63** (2013), 129–134. DOI: 10.1007/s00025-011-0177-y.

[93] O. Kowalski, Z. Vlasek, and B. Opozda, On locally nonhomogeneous pseudo-Riemannian manifolds with locally homogeneous Levi–Civita connections, *Int. J. Math.* **14** (2003), 559–572. DOI: 10.1142/s0129167x03001971.

[94] W. Kühnel and H.-B. Rademacher, Conformal geometry of gravitational plane waves, *Geom. Dedicata* **109** (2004), 175–188. DOI: 10.1007/s10711-004-2453-4.

[95] G. M. Kuz'mina, Some generalizations of the Riemann spaces of Einstein, *Math. Notes* **16** (1974), 961–963; translation from *Mat. Zametki* **16** (1974), 619–622. DOI: 10.1007/bf01104264.

[96] M.-L. Labbi, Double forms, curvature structures and the (p, q)-curvatures, *Trans. Am. Math. Soc.* **357** (2005), 3971–3992.

[97] M.-L. Labbi, On Gauss–Bonnet curvatures, SIGMA, *Symm. Integrab. Geom. Meth. Appl.* **3** (2007), Paper 118, 11 pp. DOI: 10.3842/sigma.2007.118.

[98] M.-L. Labbi, Variational properties of the Gauss–Bonnet curvatures, *Calc. Var. Partial Diff. Eq.* **32** (2008), 175–189. DOI: 10.1007/s00526-007-0135-4.

[99] F. Lastaria, Homogeneous metrics with the same curvature, *Simon Stevin* **65** (1991), 267–281.

[100] J. Lauret, Ricci soliton homogeneous nilmanifolds, *Math. Ann.* **319** (2001), 715–733. DOI: 10.1007/pl00004456.

[101] T. Leistner, Conformal holonomy of C-spaces, Ricci-flat, and Lorentzian manifolds, *Diff. Geom. Appl.*, **24** (2006), 458–478. DOI: 10.1016/j.difgeo.2006.04.008.

[102] T. Leistner and P. Nurowski, Conformal pure radiation with parallel rays, *Class. Quant. Grav.* **29** (2012), no. 5, 055007, 15pp. DOI: 10.1088/0264-9381/29/5/055007.

[103] D. Lovelock, The Einstein Tensor and its generalizations, *J. Math. Phys.* **12** (1971), 498–501. DOI: 10.1063/1.1665613.

[104] G. K. Martin and G. Thompson, Nonuniqueness of the metric in Lorentzian manifolds, *Pacific J. Math.* **158** (1993), 177–187. DOI: 10.2140/pjm.1993.158.177.

[105] G. Maschler, Almost soliton duality, *Adv. Geom.* **15** (2015), 159–166. DOI: 10.1515/advgeom-2015-0007.

[106] O. Munteanu and N. Sesum; On gradient Ricci solitons, *J. Geom. Anal.* **23** (2013), 539–561. DOI: 10.1007/s12220-011-9252-6.

[107] L. Ni and N. Wallach; On a classification of gradient shrinking solitons, *Math. Res. Lett.* **15** (2008), 941–955. DOI: 10.4310/mrl.2008.v15.n5.a9.

[108] K. Nomizu, On local and global existence of Killing vector fields, *Ann. Math.* **72** (1960), 105–120. DOI: 10.2307/1970148.

[109] K. Onda, Lorentz Ricci solitons on 3-dimensional Lie groups, *Geom. Dedicata* **147** (2010), 313–322. DOI: 10.1007/s10711-009-9456-0.

[110] B. Opozda, On locally homogeneous G-structures, *Geom. Dedicata* **73** (1998), 215–223.

[111] B. Opozda, A classification of locally homogeneous connections on 2-dimensional manifolds, *Diff. Geom. Appl.* **21** (2004), 173–198. DOI: 10.1016/j.difgeo.2004.03.005.

[112] J. H. Park, Euler–Lagrange formulas for pseudo-Kähler manifolds, *J. Geom. Phys.* **99** (2016), 239–243. DOI: 10.1016/j.geomphys.2015.10.012.

[113] V. K. Patodi, Curvature and the eigenforms of the Laplace operator, *J. Diff. Geom.* **5** (1971), 233–249. DOI: 10.4310/jdg/1214429791.

[114] V. Pecastaing, On two theorems about local automorphisms of geometric structures, *Ann. Inst. Fourier (Grenoble)* **66** (2016), 175–208. DOI: 10.5802/aif.3009.

[115] O. Pekonen, The Einstein field equation in a multidimensional universe, *Gen. Relativ. Gravit.* **20** (1988), 667–670. DOI: 10.1007/bf00758971.

[116] P. Petersen and W. Wylie, On gradient Ricci solitons with symmetry, *Proc. Amer. Math. Soc.* **137** (2009), 2085–2092. DOI: 10.1090/s0002-9939-09-09723-8.

[117] P. Petersen and W. Wylie, Rigidity of gradient Ricci solitons, *Pacific J. Math.* **241** (2009), 329–345. DOI: 10.2140/pjm.2009.241.329.

[118] P. Petersen and W. Wylie, On the classification of gradient Ricci solitons, *Geom. Topol.* **14** (2010), 2277–2300. DOI: 10.2140/gt.2010.14.2277.

[119] S. Pigola, M. Rigoli, M. Rimoldi, and A. Setti, Ricci almost solitons, *Ann. Sc. Norm. Super. Pisa Cl. Sci.* **10** (2011), 757–799.

[120] F. Podestà and A. Spiro, *Introduzione ai Gruppi di Trasformazioni*, Volume of the Preprint Series of the Mathematics Department V. Volterra of the University of Ancona, Via delle Brecce Bianche, Ancona, ITALY (1996).

[121] R. Ponge and H. Reckziegel, Twisted products in pseudo-Riemannian geometry, *Geom. Dedicata* **48** (1993), 15–25. DOI: 10.1007/bf01265674.

[122] C. Procesi and G. Schwarz, Inequalities defining orbit spaces, *Invent. Math.* **81** (1985), 539–554. DOI: 10.1007/bf01388587.

[123] F. Prüfer, F. Tricerri, and L. Vanhecke: Curvature invariants, differential operators and local homogeneity, *Trans. Amer. Math. Soc.* **348** (1996), 4643–4652. DOI: 10.1090/S0002-9947-96-01686-8.

[124] K. Sekigawa, On some 3-dimensional Riemannian manifolds, *Hokkaido Math. J.* **2** (1973), 259–270. DOI: 10.14492/hokmj/1381758986.

[125] I. M. Singer, Infinitesimally homogeneous spaces, *Comm. Pure Appl. Math.* **13** (1960), 685–697. DOI: 10.1002/cpa.3160130408.

[126] A. Spiro, A remark on locally homogeneous Riemannian spaces, *Results Math.* **24** (1993), 318–325. DOI: 10.1007/bf03322340.

[127] M. Steller, Conformal vector fields on spacetimes, *Ann. Global Anal. Geom.* **29** (2006), 293–317. DOI: 10.1007/s10455-005-9001-9.

[128] S. Sternberg, *Lectures on Differential Geometry*, Prentice-Hall, Inc., Englewood Cliffs, N.J. 1964.

[129] Z. I. Szabo, Structure theorems on Riemannian spaces satisfying $R(X,Y) \cdot R = 0$. I. The local version, *J. Diff. Geom.* **17** (1982), 531–582. DOI: 10.4310/jdg/1214437486.

[130] H. Takagi, Conformally flat Riemannian manifolds admitting a transitive group of isometries, *Tôhoku Math. J.* **27** (1975), 103–110. DOI: 10.2748/tmj/1178241040.

[131] Y. Tashiro, Complete Riemannian manifolds and some vector fields, *Trans. Amer. Math. Soc.* **117** (1965), 251–275. DOI: 10.1090/s0002-9947-1965-0174022-6.

[132] A. G. Walker, Canonical form for a Riemannian space with a parallel field of null planes, *Quart. J. Math. Oxford Ser. (2)* **1** (1950), 69–79. DOI: 10.1093/qmath/1.1.69.

[133] Q.-M. Wang, Isoparametric functions on Riemannian manifolds I, *Math. Ann.* **277** (1987), 639–646. DOI: 10.1007/bf01457863.

[134] H. Weyl, *The classical groups. Their invariants and representations*, Fifteenth printing. Princeton Landmarks in Mathematics. Princeton Paperbacks. Princeton University Press, Princeton, NJ, 1997. DOI: 10.1515/9781400883905.

[135] C. Will, The space of solvsolitons in low dimensions, *Ann. Global Anal. Geom.* **40** (2011), 291–309. DOI: 10.1007/s10455-011-9258-0.

[136] Z.-H. Zhang, Gradient shrinking solitons with vanishing Weyl tensor, *Pacific J. Math.* **242** (2009) 189–200. DOI: 10.2140/pjm.2009.242.189.

[137] S. Zhu, The classification of complete locally conformally flat manifolds of nonnegative Ricci curvature, *Pacific J. Math.* **163** (1994), 189–199. DOI: 10.2140/pjm.1994.163.189.

Authors' Biographies

ESTEBAN CALVIÑO-LOUZAO

Esteban Calviño-Louzao[1] is a member of the research group in Riemannian Geometry at the Department of Geometry and Topology of the University of Santiago de Compostela (Spain). He received his Ph.D. in 2011 under the direction of E. García-Río and R. Vázquez-Lorenzo. His research specialty is Riemannian and pseudo-Riemannian geometry. He has published more than 20 research articles and books.

EDUARDO GARCÍA-RÍO

Eduardo García-Río[2] is a Professor of Mathematics at the University of Santiago (Spain). He is a member of the editorial board of Differential Geometry and its Applications and The Journal of Geometric Analysis and leads the research group in Riemannian Geometry at the Department of Geometry and Topology of the University of Santiago de Compostela (Spain). He received his Ph.D. in 1992 from the University of Santiago under the direction of A. Bonome and L. Torrón. His research specialty is Differential Geometry. He has published more than 120 research articles and books.

[1] Dir. Xeral de Educación, Formación Profesional e Innovación Educativa, San Caetano, s/n, 15781 Santiago de Compostela, Spain.
email: `estebcl@edu.xunta.es`
[2] Department of Geometry and Topology, Faculty of Mathematics, University of Santiago de Compostela, 15782 Santiago de Compostela, Spain.
email: `eduardo.garcia.rio@usc.es`

PETER B GILKEY

Peter B Gilkey[3] is a Professor of Mathematics and a member of the Institute of Theoretical Science at the University of Oregon. He is a fellow of the American Mathematical Society and is a member of the editorial board of Results in Mathematics, Differential Geometry and its Applications, and The Journal of Geometric Analysis. He received his Ph.D. in 1972 from Harvard University under the direction of L. Nirenberg. His research specialties are Differential Geometry, Elliptic Partial Differential Equations, and Algebraic topology. He has published more than 260 research articles and books.

JEONGHYEONG PARK

JeongHyeong Park[4] is a Professor of Mathematics at Sungkyunkwan University and is an associate member of the KIAS (Korea). She received her Ph.D. in 1990 from Kanazawa University in Japan under the direction of H. Kitahara. Her research specialties are spectral geom-etry of Riemannian submersion and geometric struc-tures on manifolds like eta-Einstein manifolds and H-contact manifolds. She organized the geometry section of AMC 2013 (he Asian Mathematical Conference 2013), the ICM 2014 satellite conference on Geometric analysis, and geometric structures on manifolds (2016). She has published more than 81 research articles and books.

[3]Mathematics Department, University of Oregon, Eugene OR 97403 U.S.
email: `gilkey@uoregon.edu`
[4]Mathematics Department, Sungkyunkwan University, Suwon, 16419, Korea
email: `parkj@skku.edu`

RAMÓN VÁZQUEZ-LORENZO

Ramón Vázquez-Lorenzo[5] is a member of the research group in Riemannian Geometry at the Department of Geometry and Topology of the University of Santiago de Compostela (Spain). He is a member of the Spanish Research Network on Relativity and Gravitation. He received his Ph.D. in 1997 from the University of Santiago de Compostela under the direction of E. García-Río. His research focuses mainly on Differential Geometry with special emphasis on the study of the curvature and the algebraic properties of curvature operators in the Lorentzian and in the higher signature settings. He has published more than 50 research articles and books.

[5]Department of Geometry and Topology, Faculty of Mathematics, University of Santiago de Compostela, 15782 Santiago de Compostela, Spain.
email: `ravazlor@edu.xunta.es`

Index

Printed in the United States
by Baker & Taylor Publisher Services